Writing for Science

A practical handbook for science, engineering and technology students

Heather Silyn-Roberts

Addison Wesley Longman New Zealand Limited
46 Hillside Road, Auckland 10,
New Zealand

Associated companies throughout
the world

First published 1996

ISBN 0 582 87816 0

Produced by Addison Wesley Longman New Zealand Limited
Printed in Malaysia through Longman Malaysia, PP
Typeset in Palatino 10/12pt.

Contents

Acknowledgements

I am grateful to the following colleagues who have contributed to or commented on the material in the book: Michael Allen, Richard Christie, Adrian Croucher, Margaret Goldstone, Murray Gregory, Patsy Hulse, Richard Hunt, Peter Jackson, Gordon Mallinson, Michael Miller, Rick Mugridge, Kerry Rodgers, David Stringer and Euan Young; and Laraine Allen of Worley Consultants Ltd.

Lawrence Carter and Richard Fenwick deserve particular thanks for their insightful reading of the complete manuscript, as does Roy Sharp for the School of Engineering's support.

Many undergraduate and postgraduate students, professional engineers and scientists with whom I have worked have sharpened my thoughts with their problems and ideas. I truly value their input.

Finally, I thank Siân and Gretel Silyn-Roberts, for their perceptiveness and their invaluable perspectives as undergraduate and postgraduate students. They gave me the most rigorous feedback and the best-humoured support.

What this book does

This book's aim is to make writing more productive and less agonising for science, engineering and technology students. It does this by:

- Assuming you have no prior knowledge of the basic requirements for science essay writing nor for technical reports. Course guidelines on writing often take this basic knowledge for granted.
- Addressing the common problems and mistakes of students in their early years.
- Giving examples taken directly from student writing, together with methods for correcting them.
- Presenting the material in a practical, direct format with listed information that is readily visible and accessible.

As a science, engineering or technology student, you may hate writing. It's a common feeling, as is the hope that you've left writing behind for ever through your choice of subjects. It may have come as a nasty shock to find that faculties require a large component of writing, because good written skills are now in high demand by all types of outside organisations.

This book is the result of dealing first-hand with hundreds of students at all undergraduate and graduate levels together with professional scientists and engineers – their problems and mistakes, their particular quirks of how they like to access information and how they set about sorting things out.

My advice to any students who doubt whether they can write effectively is this: what you need for adequate science writing is logic, precision and the ability to marshal facts – exactly the characteristics you'll need in your future profession. Anything over and above that – the choice of words that is usually called style – will lift your writing from the adequate to the polished. This book is designed to help.

SECTION 1

Writing an essay

Introduction: writing an essay

Essay-writing is an area that many science and engineering students feel ill-prepared for. Yet the area is very important: essays make up a major part of a science student's coursework and final assessment. In the earth and biological sciences in particular, assessment is based almost entirely on essay-writing, especially in the final years of the degree.

If you feel uneasy about essay-writing, don't be alarmed: you're not alone. It's a characteristic of science and engineering students that they often feel that they were bad at English in school, particularly in the sorts of essays required in English, and they transfer this feeling to their science writing at tertiary level. You may also think that you don't get enough practice at it; it is very common to have to write only one or two coursework essays during the year, and then be examined almost entirely by having to write three or four essays under pressure.

What this section of the book does:

- It is aimed at removing some of the uncertainties you may have about essay-writing in science, especially if you are in your first year. It assumes you know nothing about it, and it aims to give some guidelines and strategies for approaching the process.
- It doesn't attempt a full treatment of the art of essay-writing. The Further Reading section (page 178)) gives examples of books that deal with this area.

Other aspects of the writing process that are relevant to essay writing are covered in Section 3, **The Tools of Technical Writing.**
These are:

Choosing an essay question

CHAPTER 1

This chapter covers how to choose an essay question by taking into account:

- the type of question
- the source material
- the time needed to complete it.

An essay must be to-the-point. A logically structured, well-supported, superbly presented essay that misses the point is a poor essay. It is all too easy to lose marks by writing an essay that you think is relevant, but which does not address the main focus of the question.

Choosing an essay question

Obviously, you will choose a topic that already interests you or that you think will be interesting after you have read around it. But there are two other factors that should influence your choice:

1. **Which type of essay do you want to write?**
   Although the range of essay questions is vast, they can be roughly classified into two types:
   - those that require you to **describe**
   - those that require you to **present an argument**.

   Chapter 2 **Analysing an essay question** will help you to recognise from the wording of the question the type of essay that is required.

2. **Is there enough source material?**
   Choose a topic for which there is a ready supply of source material – books, articles, pamphlets, publicity material, newspaper articles etc. Be aware that it is possible to spend a considerable amount of time looking for material, and to leave yourself too little time if you decide to change your topic because of lack of resources.

How much time is needed?

If you can choose your question, make your choice early. This will help you to:

- get to the material in the library before the rush (see Chapter 10 **Doing the background reading)**
- mull over the question in your mind; this often generates new insights
- keep a look out for additional material like newspaper cuttings, TV programmes etc.

Factors to keep in mind here are:

1. **Writing something – essays, reports, anything – always takes much longer than you think.** Allow far more time than you think you will need. It is always amazing how much time the final stages take; many people are often caught unawares.

2. **Essay-writing needs time management.** An essay is a piece of investigation. It can't be dashed off in a couple of hours or a day. It involves planning, reading, collating information, writing, revising, editing, proofreading, and a lot of thinking. Enter times for these on your weekly timetable. You may not keep to it rigidly, but it will help you to keep control of the process.

3. **Be aware of what your lecturers expect.** If you are given a long period between the setting of the essay and its submission date, they will expect you to show that you have taken the time to research the topic.

Checklist: Choosing an essay question

Ask:

- ✔ Which type of essay question do I want to write? Descriptive or interpretive/argumentative?
- ✔ Is there enough source material?
- ✔ How much time do I need? Allow for a lot more than you would expect.

Analysing an essay question

2

CHAPTER

This chapter covers:

- recognising and knowing the meaning of the key instruction words used in essay questions
- giving the required focus to your essay.

It is vital to analyse an essay question accurately. It establishes the focus for your answer, which you then keep in the forefront of your mind whilst writing the essay. Many marks have been lost through inaccurately analysing the question at the beginning, or drifting away from it after an initial sharp focus.

An essay needs an answer to a specific instruction or question. There are enormous differences between being asked to describe something, or to evaluate, or analyse, or summarise something. The key to establishing the focus for your essay is being able to identify the key instruction words and their meaning.

This chapter explains:

- the meanings of the key words and phrases used
- how to take apart an essay question to establish the focus.

What are the key words?

The key words most often found in undergraduate science-based essays and their brief meanings are given below. The next section groups these words into themes, and gives expanded meanings, examples of essays and essay plans.

Account for	Give the details of, give the reasons or the underlying cause of
Analyse	Examine closely, examine *x* in terms of its components and show how they interrelate
Assess	Give a thorough, balanced evaluation or account of
Compare and contrast	Discuss *x* and *y* in terms of their similarities and differences

Compare	Discuss *x* and *y*, emphasising their similarities
Contrast	Discuss *x* and *y*, emphasising their differences
Consider	Give a thorough, balanced evaluation or account of
Define	Explain what is meant by, give a definition
Demonstrate	Give examples to: • show how/that • explain a concept
Describe	Give an account of
Discuss	Give a thorough, balanced evaluation or account of
Evaluate	Try to form a judgement about
Examine	Give a thorough, balanced evaluation or account of
Explain	Give the details of, give the reasons or the underlying cause of
Identify	Highlight the important characteristics or reasons
Illustrate	Give examples to: • show how/that • explain a concept
Indicate	Highlight the important characteristics or reasons
List	Highlight the important characteristics or reasons
Outline	Highlight the important characteristics or reasons
What?/how?/ why?/ to what extent?	Give an account of
Write an essay on..., write an account of...	Give an account of

Which type of essay question is it?

The key words give the clues as to which type of essay is required.

The essay that asks you to describe	The essay that asks you to present an argument	The essay that is part description, part argument
explain	assess	analyse
demonstrate	evaluate	compare
describe	examine	compare and contrast
identify		consider
outline		contrast
summarise		discuss

How to analyse an essay question

The key words in an essay question form the cornerstone to writing a relevant answer. The first thing that must be done in assessing an essay question is to analyse the wording of the question, in order to identify the key instruction words and the key topic words.

Here are suggested steps in analysing an essay question:

Step 1: Circle the **key instruction word(s)** in the question – describe, evaluate, discuss, outline etc.

> Discuss the development of manufacturing industries in China since 1949.

Check that you understand what it implies (use the next section to help you do this.) Remember to analyse the *whole* essay question this way. Many essay questions contain two or more keywords in separate parts of the question, e.g.

> Discuss the application of sustainability theories to either horticulture, or inshore fisheries or native timber harvesting. Evaluate the progress of current moves towards sustainability in these industries in this country. Include a discussion of international directives and this country's statutes.

In these multi-part questions, make sure that you analyse each part thoroughly.

Step 2: Underline the **key topic words** that define the boundaries of your essay.

> Discuss the development of manufacturing industries in China since 1949

> *You have chosen inshore fisheries:* Discuss the application of sustainability theories to either horticulture, or inshore fisheries or native timber harvesting. Evaluate the progress of current moves towards sustainability in these industries in this country. Include a discussion of international directives and this country's statutes.

Step 3: Identify the **focus** of the question. Think about which aspect of the topic you are being asked to write about.

> Discuss the ***development*** over time of manufacturing industries in China since 1949.

> Discuss the *application* of sustainability theories to either horticulture, or inshore fisheries or native timber harvesting. Evaluate the *progress of current moves* towards sustainability in these industries in this

country. Include a discussion of international directives and this country's statutes.

Step 4: Brainstorm a **series of questions about the topic**. These give you a useful starting point for your research.

For the first example:
- What characterised manufacturing industry in China before 1949?
- What characterises it now?
- What stages of development did it go through?
- What influenced the process?

For the second example:
- What sustainability theories can be applied to inshore fisheries?
- How can they be applied?
- What are the international directives in this respect?
- What are the current moves towards sustainability in this country?
- What are this country's statutes in this respect?
- How are the moves progressing?

Reasons for going through these steps:

Going through the steps outlined above will do three things:

1. It will ensure that your essay has the **focus that your marker requires.**
2. By having a clear focus, it will ensure that the **time you spend reading for your topic is well-spent.** This way you don't end up with a formless mass of material. Without a clear focus, it is easy to read too widely.
3. It will ensure that **you don't become overwhelmed** by the sheer amount of information in the field.

What do the key words mean and how are they used?

The next section describes the requirements for the most common key instruction words used in science-based essay questions. Given with each set of key words are:

- their meanings
- examples of essay topics using these words
- a model of an essay structure using these key words.

Discuss

Meaning

- Give a thorough, balanced evaluation or account of.
- Present the different aspects of a subject.
- Investigate by reasoning or argument.

This key word often occurs as the second part of an essay question, particularly in exams (see pages 14-15)

Examples

1. Discuss the application of sustainability theories to either horticulture, or inshore fisheries or native timber harvesting. Evaluate the progress of current moves towards sustainability in these industries in this country. Include a discussion of international directives and this country's statutes.
2. In the Great Barrier reef, the composition of various coral reef communities changes along a gradient from inshore coastal to offshore oceanic habitats. Discuss the physical and biological factors that may create and maintain such patterns.
3. Discuss the land-sea boundary with specific reference to the dominant processes in estuaries.
4. One of the primary goals in biogeography is to explain why plant and animal species live where they do. Discuss how the study of spatial patterns of species distributions can be used in biological conservation. Use specific case studies in your discussion.
5. Discuss the development of manufacturing industries in China since 1949.
6. Neither the developed nor the developing countries of the world can afford to reduce their future carbon dioxide emissions. Discuss, giving specific examples.

Possible essay structure

(see also note below):

1 Introduction: give a brief general overview of the topic and its significance.

2 **a** If there is no statement at the beginning of the question:
- give a thorough, balanced account of the topic
- for the Conclusion summarise (i) the most important features (ii) the main pieces of evidence.

b If there is a statement (see question 6 above):
- describe and analyse it
- give the evidence that (i) supports the statement and (ii) runs counter to it
- for the Conclusion (i) summarise the evidence that is most strongly for and against the statement and (ii) state your conclusion about the extent to which the statement is justified.

Note: The difference between Arts and Science in defining essay focus. *Discuss* in an arts or social science essay usually means giving evidence that both supports a statement or viewpoint and runs counter to it. The evidence is presented as a reasoned argument, often with your own views expressed. However, in an undergraduate science essay – particularly in an exam – *discuss* often requires you to give a **thorough, balanced account of an area of knowledge** (see most of the previous examples). You need to be aware of this subtlety.

Evaluate, assess, examine, consider

Meaning

- Give a thorough, balanced account.
- Inspect closely.
- Estimate the value or significance.

Examples

1. Evaluate the role of biodiversity in the maintenance of ecosystem function and flexibility. Use marine reserves and marine resource utilisation as examples in your discussion.
2. The world must soon begin to shift its energy base away from non-renewable fossil fuels to create a greater reliance on renewable energy resources. Evaluate the application of precautionary principles within this transition.
3. Evaluate the effectiveness of the 'one-child-policy' as a means of population control in China.
4. Security in the city is a function of social problems. Examine this assertion with reference to the association of increasing crime with extended urbanisation.

Possible essay structure

1. Introduction: give a brief general overview of the topic and its significance.
2. Break the topic down into its various aspects.
3. Give evidence relating to each of its aspects and to the links between them.
4. At the same time, come to conclusions about how valid each aspect is.
5. Conclusion: (i) summarise the main components and the related evidence and (ii) give brief conclusions about the validity of the main ideas.

Compare and contrast, compare, contrast

Meaning

- *Compare and contrast, compare:* discuss x and y in terms of their similarities and differences.
- *Contrast:* discuss x and y, emphasising their differences.
- What important features do x and y have in common?
- In what important ways is each one unique?

Examples

1 Compare and contrast the bryophytes with the ferns and their allies.
2 Compare and contrast modernisation processes and the contemporary characteristics of manufacturing industries in China and Japan.
3 Compare and contrast the sediment textures and sedimentary structures of typical coastal/shallow marine and bathal/deep marine environments.
4 Compare and contrast the new (household) responsibility system and the People's Commune system in Chinese agriculture.
5 Contrast the activation of contraction of cardiac and smooth muscle.

Possible essay structure

1 Introduction: briefly describe the two or more things you have been asked to compare and contrast, and the significance of each.
2 Analyse each one for its components, describing similarities, or differences or both.
3 Summarise, giving an overview of the most important ways in which the two things are similar and / or different.
4 You will probably not spend the first half of your essay on *x* and the second half on *y*. It is better to introduce each individual component in turn, relating each one first to *x* and then to *y*.

Describe, identify, list, what?, which?, how?, to what extent?

Meaning

- Give an account – clear, well-organised, main points emphasised.
- Present in turn each of the main characteristics of the phenomenon.

These key words are often used in the first part of an essay question, particularly in exams (see pages 14-15).

Examples

1 Describe the geometrical principles used to determine the rate of convergence between two plates at a trench.
2 Estuaries are effecient sediment traps. With reference to wave-dominated estuaries, describe the evolution of an estuary from the time of formation to the time of complete filling.
3 Describe the process of DNA replication as it occurs in bacteria.
4 Identify the properties of clay which are important in ecosystem nutrient cycling.
5 Identify the major external and internal forces contributing to the growth of manufacturing in the countries of South-East Asia.
6 Why is the study of plate tectonics relevant to an understanding of the global distribution of volcanoes?
7 List the main stages in the physical development of planet earth.

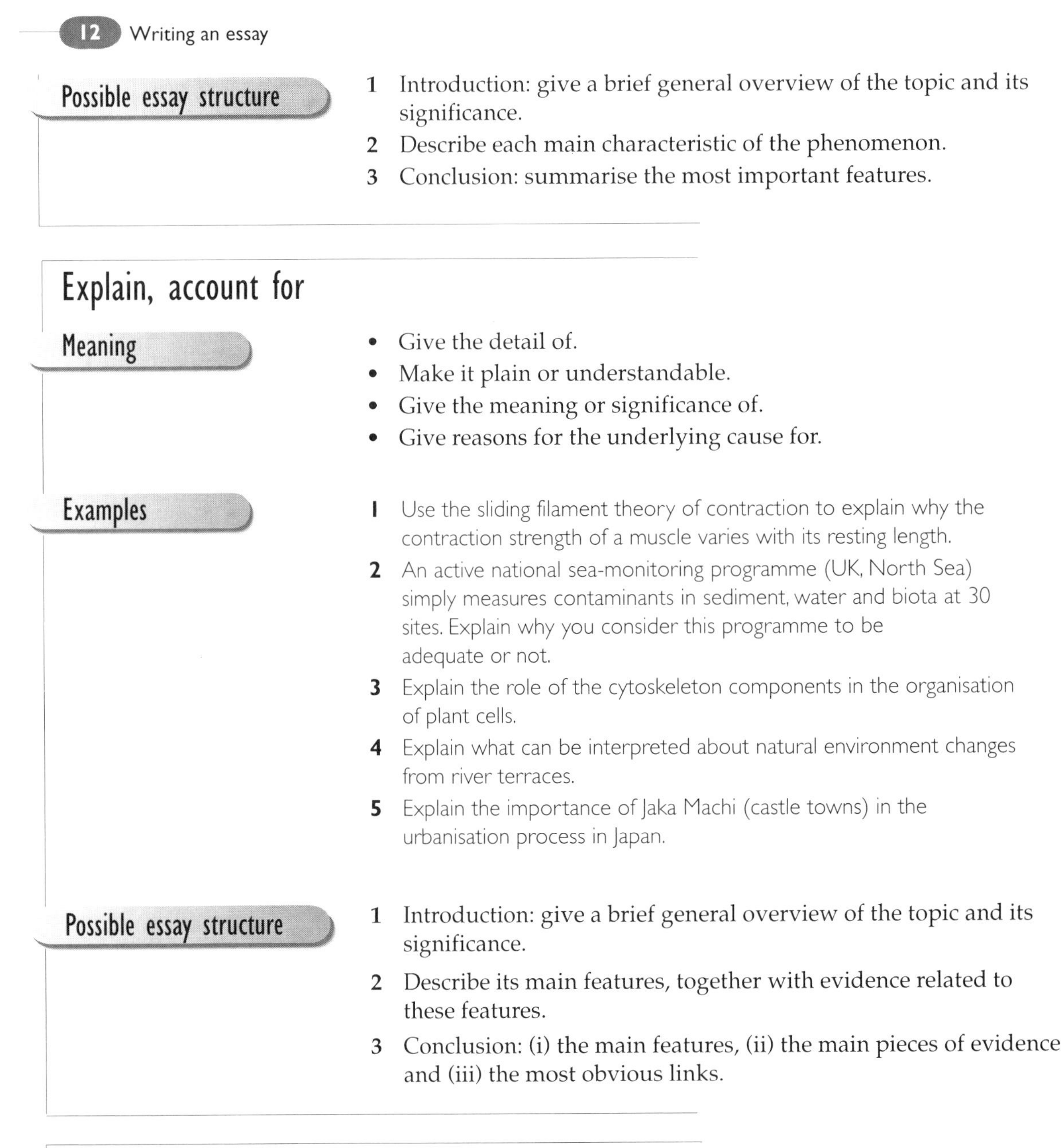

Possible essay structure

1 Introduction: give a brief general overview of the topic and its significance.
2 Describe each main characteristic of the phenomenon.
3 Conclusion: summarise the most important features.

Explain, account for

Meaning

- Give the detail of.
- Make it plain or understandable.
- Give the meaning or significance of.
- Give reasons for the underlying cause for.

Examples

1 Use the sliding filament theory of contraction to explain why the contraction strength of a muscle varies with its resting length.
2 An active national sea-monitoring programme (UK, North Sea) simply measures contaminants in sediment, water and biota at 30 sites. Explain why you consider this programme to be adequate or not.
3 Explain the role of the cytoskeleton components in the organisation of plant cells.
4 Explain what can be interpreted about natural environment changes from river terraces.
5 Explain the importance of Jaka Machi (castle towns) in the urbanisation process in Japan.

Possible essay structure

1 Introduction: give a brief general overview of the topic and its significance.
2 Describe its main features, together with evidence related to these features.
3 Conclusion: (i) the main features, (ii) the main pieces of evidence and (iii) the most obvious links.

Illustrate, give examples of, demonstrate, use a specific case study to

Meaning

- Give examples to:
  a show how/that
  b explain a concept.
- Make clear by explanations.
- In some contexts, *illustrate* can mean draw diagrams. However, in essay questions it is almost always an instruction to give examples.

Examples

1 Are kelp forest dynamics controlled primarily by sea-urchin grazing or by episodic, climate-related processes? Use detailed examples from published studies to illustrate your answer.
2 Describe the three types of predator response to variations in prey density. How are they different, and why? Illustrate your answers.
3 List the attributes of an 'ideal' index of human thermal climate. Give examples of bioclimate indices that have these attributes.

Possible essay structure

1 Introduction: give a brief general overview of the topic and its significance.
2 Give evidence and examples that support the phenomenon, and how the phenomenon gives rise to appropriate concepts.
3 Conclusion: summarise the main links between the phenomenon and the main supporting examples.

Outline, summarise, indicate

Meaning

- Give an overview of the topic.
- Extract the main points.
- Be clear, concise and follow a logical sequence.
- Often used in the first part of an essay question (see pages 14-15).

Examples

1 Members of the Crustacea utilise almost every zone of the lake ecosystem. Briefly outline the major zones and the Crustacea which occupy them.
2 Using one South Pacific island country as a case study, outline the way the economy was transformed following European contact.
3 Outline the primary controls in the development of karst landforms.
4 Outline the field, laboratory and analysis procedures you would use to examine the development and the maintenance of a small river and flood plain system.
5 How are sugars taken up into isolated cells by fluid-phase endocytosis (pinocytosis)? Indicate some of the experimental difficulties in measuring the uptake.

Possible essay structure

1 Introduction: give a brief general overview of the topic and its significance.
2 Analyse it, emphasising only the main features of the topic.
3 Conclusion: summarise giving a statement that focuses on the most important features.

Define

Meaning

- Give a definition.
- Formulate precise statements of meaning.
- Sometimes used in the first part of essay questions (see page 00).

Examples

1 Define micro-, meso- and macro-tidal coasts and discuss the influence mean tidal range has upon coastal depositional landforms.

Possible essay structure

1 Introduction: give a brief general overview of the topic and its significance.
2 Give definitions and/or clear descriptions of the components.
3 *Define* is usually in combination with *discuss* or another similar keyword.
4 Concusion: summarise the most important features.

Combinations of key words

Many essay questions use two key information words:

- the first one (usually *describe, identify, list, outline, explain, define*) expects you to give a clear, concise description of something
- the second one (nearly always *discuss*) expects you to give a thorough, balanced account of it in relation to a given phenomenon.

These essay questions illustrate this:

- *Describe* and briefly *discuss* the processes which are active during metamorphic recrystallisation.
- *Describe* the phenomenon of referred pain and *discuss* the neural mechanisms that subserve it.
- *Describe* the process of evapotranspiration and the major factors controlling this process. Then *discuss* why one might expect higher rates of evaporation from an area of forest than from an area of pasture.
- *Identify* and *discuss* the outcomes of European colonial expansion.
- *List* the main effects that light has on the control of plant growth. *Discuss* the extent to which flowering is controlled by light.
- *Outline* the Huxley (1957) theory of contraction. *Discuss* the extent to which this model is consistent with the structure, and the energetic and mechanical behaviour of skeletal muscle.
- *Explain* where water is stored in a river basin, how the stores are linked and *discuss* what controls the residence time of water in each store.

This is a very common format for an essay question, particularly in exams. Make sure that you answer the question in that order: first describe, then discuss. This format ensures a logically structured answer. The basic description or listing makes a good introduction to the essay and gives you something to build on for the account in the main text of the essay.

Now that you have analysed the essay question, use Chapter 10 **Doing the background reading** for help with the next stage.

Checklist: Analysing an essay question

- Analyse the key instruction word(s) – *discuss, compare and contrast* etc. – and establish what each one needs.
- Find the key topic words that define the boundary of the essay.
- Identify the focus of the question.
- Brainstorm a series of questions about the topic.
- Then do the background reading (see Chapter 10 **Doing the background reading**).

CHAPTER 3

Drawing up a plan and writing the first draft

This chapter covers:

- constructing a detailed plan for your essay
- incorporating the material from your background reading into the essay plan
- writing the rough draft of the essay.

First use Chapter 10 **Doing the background reading**. Then use this chapter to plan and write the first draft.

Why plan?

- **The writing has a more logical structure**. Spontaneous writing usually produces a 'stream-of-consciousness' or a memory dump – the essay is without direction, and your reader can sense your lack of focus.
- **Less time is ultimately needed**. The more time that is spent in the planning phase, the less time is spent in writing, because unplanned material needs to be revised more.

Where do I start?

Keep in mind your analysis of the key instruction words and the key topic words in your essay question. The first step is then to decide which of the following essay schemes is appropriate.

What type of essay scheme do I need?

There are two basic types of essay:

1 The descriptive essay

These require you to inform the reader by giving a coherent and balanced account of your subject matter.

- Very many essays at the early undergraduate level in the sciences are descriptive.
- Exam essays at all levels are often descriptive.

2 The argumentative or interpretive essay

- You are less likely in the early years of a science degree to be asked to write the type of essay that requires your personal interpretation of an abstract idea e.g.

  'Sustainable development is good in theory, but is difficult to put into practice.' Debate, using specific examples.

- But many essays will ask you to select and evaluate information, e.g.

  Evaluate this country's governmental response to climate change with particular reference to carbon dioxide and methane emissions.

Such essays will require a balance of the characteristics of both descriptive and interpretive forms.

The descriptive essay

What is needed in a descriptive essay?

You need to show that you can:
- Understand the issues clearly and:
  - **a** identify the themes
  - **b** rank the ideas in importance
  - **c** understand the relationships.
- Present appropriate material, so that coverage is thorough and balanced.
- Organise this material into a logical structure with a clear focus.
- Write a clear, straightforward, concise answer.

Schemes for writing a descriptive essay

Even descriptive essays can be garbled, if they are not planned well. This section gives a number of schemes for planning this type of essay.

1 Generality schemes

Here you are asked to present a body of conceptual material, as for example where you are asked to describe a principle or theory. A lucid account is needed.

> Explain the technology of powder diffraction and how it may be applied to the study of minerals.

> Describe the current state of knowledge of the nature and actions of soluble mediator(s) of the action of insulin.

2 Time-based schemes

Ordering by time is readily understood by the reader. It is worth using if at all possible. Some topics are obviously time-based:

> Discuss the development of manufacturing industries in China since 1949.

> Describe the life cycle of the organism causing schistosomiasis.

> List the main stages in the development of planet earth.

Others can be developed as a time-based scheme:

> Estuaries are efficient sediment traps. With reference to wave-dominated estuaries, describe the evolution of an estuary from the time of formation to the time of complete filling.

> Using one South Pacific island country as a case study, outline the way the economy was transformed following European contact.

3 Process-based schemes

Some essay topics need you to trace through a process step-by-step, as you would in a flow-chart.

> Describe the process of DNA replication as it occurs in bacteria.

4 Cause-and-effect schemes

First decide on the relevant causes. Then couple these with their particular consequences.

> In the Great Barrier Reef, the composition of various coral reef communities changes along a gradient from inshore coastal to offshore oceanic habitats. Discuss the physical and biological factors that may create and maintain such patterns.

5 Importance schemes

The best way to approach this is to begin with the most important or common concept, and move from there to the secondary concepts.

> Outline and explain the parameters that are important in describing the dispersion of a plume of pollutant downstream from its point source.

6 Comparative schemes

This is usually suggested by the wording of the topic; look for the words *compare, contrast* or both.

Compare and contrast means deal with similarities and differences:

Compare and contrast the sediment textures and sedimentary structures of typical coastal/shallow marine and bathal/deep marine environments.

Compare and contrast modernisation processes and the contemporary characteristics of manufacturing industries in China and Japan.

Where you have to compare *x* with *y* (for example China and Japan in the second example), it does not mean that you should necessarily spend the first half of your essay on *x* and the second half on *y*. When you have a set of points on which to compare and contrast your subjects, the best scheme may be to introduce each point individually, relating it first to *x* and then to *y*.

Contrast requires you to deal only with the differences:

Contrast the activation of contraction of cardiac and smooth muscle.

7 Problem-solution schemes

The topic normally focuses on the solving of a problem. Tackle this by first outlining the problem, which then leads you into the solutions.

Environmentalists and others are predicting a substantial rise in sea level over the next century. Review the possible consequences of this and how coastal zone engineers and planners could respond.

Outline the current debate between Maori people and the New Zealand Department of Conservation over the kiore (Polynesian rat). What are the differences between the Maori and Western perspectives on rats? How might the debate over rats be resolved to the satisfaction of both Maori people and the Department of Conservation?

8 How and why schemes

How questions are concerned with the means by which things are achieved:

How do animals and plants measure daylength? What is the relevance of this to the timing of annual rhythmicity?

Why deals with the underlying reasons:

An active national sea-monitoring programme (UK, North Sea) simply measures contaminants in sediment, water and biota at 30 sites. Explain why you consider this programme to be adequate or not.

The interpretive essay

What is needed in an interpretive essay?

You need to show that you can:
- form and articulate a sharply-focussed evaluation or argument
- support it with factual evidence and sound logic
- anticipate objections to your evaluation or argument, and answer them
- be objective and critical.

The art of argument and interpretation is not covered in detail here. Use the Further Reading section, p 178 to find more general essay-writing books that can help. However, points to note about the interpretive essay are:

1 Choose your criteria wisely

The criteria that you choose to support your interpretation will form the basis of the structure of your essay. You cannot choose everything in the subject area. Read around the topic, then concentrate on aspects that are:
- **Important.** Do not write an essay around minor aspects.
- **Interrelated.** A well-constructed essay is one in which information is linked into a coherent whole.
- **Capable of being balanced.** The demands of the various aspects are often conflicting. Your task is to present a balanced account.

2 Avoid an emotional interpretation

Some topics where you may hold strong views could look to you as though you are being asked to present your personal viewpoint:

> 'Recycling is a sham dreamed up by an advertising executive to encourage consumers to buy over-packaged goods'. Discuss the components and structure of a comprehensive solid waste-management scheme for your region and evaluate the role of recycling within it.

> The 'precautionary approach' has been adopted by a number of groups in order to safeguard environmental systems from degradation. Critically appraise the advantages and disadvantages of this approach to the development and application of new agrochemicals.

The wording shows that you are being asked for a balanced account of a phenomenon. Remember that your lecturers are asking for your objective views as a scientist; don't go headlong into an outpouring of personal feelings. See page 143 for ways of avoiding phrases such as 'I believe', 'I feel' etc.

Drawing up an essay plan

Planning an essay should be done as a series of stages, at each of which the plan becomes more detailed.

Get something down on paper. Not all writing begins with a scheme. Some people find it impossible to work this way. A process that can help is what is sometimes called **writer-based prose.** This doesn't aim to project ideas to the reader; it is merely to get anything down on paper. By writing and harnessing your thoughts, you will suddenly find that you recognise what the assignment has to do. Then **throw that page away** and use the points that emerged to plan the headings under which the assignment can take shape.

Stage 1: work out a very broad plan for your essay topic

- You will have had a rough plan to direct your background reading, which you have now done (see Chapter 10 **Doing the background reading**). You are now ready to start planning your essay. You will have gained new ideas and perspectives through your reading, and as a result, your new essay plan may be somewhat different from your original rough plan.
- Do not worry about a detailed structure at this point; your broad plan is just a tool to help you get your notes and ideas in the right order.
- Decide which essay scheme you are going to follow.
- Brainstorm your ideas onto paper and draw up a concept map, laying out the main ideas and relationships between them. An example of a concept map is given on page 41.
- From the concept map draw up a scheme showing linear development: either a structured list or a flow chart.

Stage 2: look through all the notes you made during your reading, classifying them into these few subject areas

- Get all your material together and place it roughly in the order of the sections in your broad plan. A useful strategy is to spread your notes around on the floor, chopping them up and ordering them. You can stick small pieces of paper onto large ones, in a logical flow of ideas.
- While you are doing this, you will get a good idea of the amount of material you have in each category, and the relative importance of the different points.
- Don't worry at this stage about how to fit all the items into a detailed structure.

Stage 3: scan through all the material in one broad section and work out the detailed structure of that area

Being critically selective about your material at this stage is vital, because it is a mark of your understanding of the topic.

- Decide which of the material contributes in a positive way towards the logical flow of each section. Put this in one pile.

- Critically examine the remaining material. If it is marginal, keep it for possible inclusion.
- Otherwise, leave it out. You are bound to have picked up material along the way that later turns out to be of no use.
- Work out the detailed structure of that section, again devising a linear plan.

Stage 4: write the detailed plan

- Write it on one side of the paper, so that you can see the overall plan easily while you are writing. Keep in the forefront of your mind:
  - a the focus of your essay, as determined from the key instruction and topic words (Chapter 2 **Analysing an essay question)**
  - b the category of writing scheme (pages 16-20) that you intend to follow.
- The plan should be made up of broad headings, each of them in the logical order needed to tell the overall story. Each broad section should be subdivided into a series of sub-sections. The number of broad headings, and the extent to which they are subdivided, will depend on the scale of the essay.
- Plan for a smooth transition from one idea to the next.
- Note any interconnections you may want to make between the various sections.
- In this detailed plan, note the examples you intend to use in each sub-section. To do this, you need to read your notes again and cross-refer them to the plan. In your notes, mark in the margins the parts that are relevant to the various sections of your plan. If a set of notes has items that are relevant to more than one section of the plan, you can either (1) cross-correlate your plan and margin notes by colour-coding them, or (2) you can chop up your notes.

This is probably not the time to plan your introductory and concluding paragraphs in anything other than broad detail. You will have a far better idea of what is needed after you have written the first draft of the essay.

Writing the first draft

An effective way of writing the rough draft is to deal with each section in turn, and convert each of the main ideas in it into a block of writing. By breaking the writing down into manageable chunks, you avoid as far as possible the initial feeling that it is a frighteningly formidable task.

1 **Start with the easiest section.** If you know you are going to find one particular section easy to write, start with that section. You can find ways of blending it into the essay structure later.

2 **Make a note at each relevant point in the essay of each of the references you are going to cite.** At this point it doesn't matter whether you are required to use the author/date (Harvard) system or the sequential numbering system for citing your references (see Chapter 13 **References,** for a full explanation of referencing). The details of the method can be sorted out in the rewriting stage (see Chapter 14 **Revising the first draft**). In these early stages it's important only that you make a note of which ones you are going to use and where they are going to be placed.

3 **Decide on the illustrations you are going to use.**

4 **Do not expect every sentence to appear on the page in perfect style and completeness.** The essential thing is to write the ideas down. Clarifying them is done in the next stage: revising (see Chapter 14 **Revising the first draft**).

5 **Writer's block.** Once you have started writing, the process generates its own momentum. But do not expect the rate to be steady. Sometimes a page can be easy to write; at other times progress through a few sentences is grindingly slow. Even experienced writers have to cope with this. If you have particularly strong problems with one area, make notes of what you want to say, then move onto the next section and return to the troublesome patch later.

6 **Aim to get the basic, completed form of the essay onto paper.** Only when you have a first rough draft on paper can you see how to sort it out. Use Chapter 14 **Revising the first draft** to help you though the rewriting process.

7 **Writing by hand.** If you are handwriting the first draft, some ideas that may be helpful are:

- Write on every other line. This gives you room for corrections.
- Use one side of the paper only. This allows you to do a cut-and-paste later.
- Use a pencil. Your writing feels less scratchy.

The Introduction and the Conclusion

Probably the best time to write both of these paragraphs is when you have completed the first draft of the body of the essay.

The **Introduction** should:

- Point out the relevance of the issue. Sometimes brief background details should be given.
- State the main issues to be considered. Moreover, state the aspects you are not going to cover, and why. This prevents the assertion that you neglected to address them.

- State how you are going to cover the issues. This prepares readers by giving them a map of what to expect.

The **Conclusion** should:

- Sum up your main points.
- Present again, very briefly, your main conclusion or argument.
- Not introduce any new material.
- Not necessarily start with 'In conclusion, ...' Be confident that the material itself will indicate the role of this final paragraph.

✎ Checklist: drawing up a plan and writing the first draft

Decide which type of essay it is:

- ✓ descriptive
- ✓ argumentative/interpretive
- ✓ part description, part argument.

- ✓ If it is a **descriptive** essay: establish the type of scheme needed.
- ✓ If it is **interpretive**:
  - ✓ choose your criteria wisely
  - ✓ avoid an emotional interpretation.

Draw up the plan

- ✓ Work out a broad plan.
- ✓ Classify your notes.
- ✓ Work out the detailed structure of each section of the essay.
- ✓ Draw together all the section plans into an overall detailed plan.

Write the first draft

- ✓ Write the easiest section first.
- ✓ Accept that the ideas and sentences are not going to be perfect at this stage.
- ✓ Note which references you are going to use and where.
- ✓ Decide on your illustrations.
- ✓ Accept that writer's block is normal and have strategies for dealing with it.
- ✓ Write the Introduction and Conclusion paragraphs.

How coursework essays are marked

4

CHAPTER

This chapter covers:

- How to recognise what your marker is looking for in a coursework essay.

There are seven aspects of essay-writing that lecturers expect you to benefit from and display in your assignments. They will be looking for evidence of these aspects when they mark. The seven things are shown below arranged in a hierarchy of importance – from the most important to the least important. The second column shows the outcome when there is a lack of understanding of these concepts.

Aspects of essay writing in order of importance	Outcome as a result of lack of understanding of each concept
1 **Understanding the question properly, and its implications**	• Completely misunderstanding the question. Writing the wrong answer to the topic. Shows a very serious lack of comprehension
2 **Analysing a question correctly** Answering in line with the wording of the question	• Misunderstanding the meanings of the key words. You might be *describing* instead of *evaluating*, or *illustrating* instead of *debating*. (See Chapter 2 **Analysing an essay question**)
3 **Organising the essay into a coherent overall structure** • Identifying themes • Ranking the concepts in importance • Understanding the relationships • Organising the information into a coherent structure with a sharp focus	• Inadequate grasp of the important concepts • No logical, consistent thread of theme or argument – information is in a jumbled order – a confusing essay that is difficult to read • Including material that is not relevant to the theme. (See Chapter 3 **Drawing up a plan and writing a rough draft**)
4 **Using skills of description, evaluation, discussion and argument** • For a descriptive essay: presenting a balanced account of an area of information • For an essay that presents an argument: the argument is objective and balanced • Using appropriate illustrations	• Not supporting your theme or argument with factual evidence and sound logic • Presenting an unbalanced argument – not being objective and critical • Not anticipating objections to the argument and not answering them • Inappropriate illustrations, poorly labelled and referred to (See Chapter 12 **Illustrations**)
5 **Using a wide range of appropriate material** • Using the library efficiently • Being able to focus your reading in a particular area • Being able to sort and sift information	• Lack of knowledge. Not using material from a wide enough range of sources. (See Chapter 10 **Doing the background reading**)

Aspects of essay writing in order of importance	Outcome as a result of lack of understanding of each concept
6 Communicating precisely • Being able to write a clear, straightforward, intelligent answer that is grammatically correct • Writing interestingly • Writing legibly, if hand-writing • Using the terminology appropriate to the discipline	• Incorrect sentence structure (sentence fragments) • Faulty grammar, spelling and use of words • Unlinked sentences • Dull writing style • Untidy or illegible handwriting • Not using the appropriate specialist terms. (See Chapter 15 **Problems of style**)
7 Using the accepted editorial conventions • Correctly referring to sources in the text, and cross-referring to the References section • Structuring the References section correctly • Obeying the conventions for illustrations	• Presenting references without attention to the conventions. (See Chapter 13 **References**)

Getting feedback from a marking schedule

Some departments will give you back a marking schedule with your essay. The aim of this is to give you feedback on how your essay has measured up to the criteria in the table above. The criteria that are likely to be used in a simple marking schedule, again arranged in a hierarchy of importance, are:

- interpretation of the question
- essay structure and ordering of the argument
- sources of the information
- grasp of the content
- quality of argument
- use of diagrams
- quality of language
- referencing
- presentation.

When a schedule such as this is returned with your essay, it is very important to use it as a learning tool. Writing an essay shouldn't be seen as an unpleasant chore, and the marking schedule merely glanced at if you've gained a passing grade. Use the information contained in it to improve your essay-writing skills. Writing better essays means a better understanding of the topics. However, this can only happen if you look on the process of producing an essay, and reacting positively to the feedback, as an active process of enquiry. If you feel that you need even more information than is contained in the marking schedule, ask your marker.

Feedback when there is no marking schedule: comments only

Where marking schedules are not used by departments, feedback in the form of brief comments or ticks and crosses can be inadequate. Rather than give meaningful feedback about fundamentally important aspects of an essay, it is much easier for a marker to comment on and correct the less important ones, such as:

- occasional very minor points of grammar (such as a split infinitive)
- not citing a particular reference
- minor conventions of formatting the References section, such as missing punctuation marks.

Other aspects of grading that can be unsatisfactory are:

1 Unspecific general comments, such as 'A reasonable effort', or 'Difficult to follow'.

2 A low grade given for an essay in which a very good effort has been made to construct a valid interpretation of a series of difficult concepts, but which is untidily presented.

3 A high grade given to a beautifully-presented essay that:
   - faithfully reproduces, word-for-word, sources that are often uncited
   - accurately paraphrases the set text or lecture notes but which shows no real understanding of the concepts.

What to do if you think the feedback on your essay is inadequate

If you have an essay returned with a low grade and unhelpful comments, then you should not be afraid to ask the marker the reasons behind the low grading, and how you can improve your skills in readiness for the next essay. This is a fundamental part of learning, and any marker should be prepared to talk to you about it.

Checklist: how coursework essays are marked

The main features that a marker will look for, in order of importance

- ✓ Understanding the question.
- ✓ Analysing the question correctly.
- ✓ Organising the essay into a coherent overall structure.
- ✓ Using skills of description, evaluation, discussion and argument.
- ✓ Using a range of appropriate material.
- ✓ Communicating precisely.
- ✓ Using the accepted editorial conventions.

After the essay is returned to you, use the feedback as a learning tool

- ✓ Get feedback:
  - ✓ from the marking schedule
  - ✓ if no schedule, from the comments
  - ✓ if comments are inadequate, ask the marker for feedback.

Writing under pressure: strategies for writing essays in exams

5

CHAPTER

This chapter covers:

- developing useful strategies for determining possible exam essay questions
- swotting productively for essay-writing in exams
- writing essays under pressure of time.

It is not unusual for many people to feel panicky about essay writing in exams because they know they haven't had enough practice during the year. Yet essays make up the main component of the final exams in many subjects.

This chapter will cover the main strategies for writing essays under exam conditions.

Try to identify possible exam topics

- Study the course outline to work out how the lecturer views the structure of the course.
- Look at previous exam papers.

By these two methods, you can often detect broad themes that are likely to come up as exam topics. Many lecturers have favourites. But you need to check if the lecturer or course outline has changed recently. If there's been a change, don't despair – each topic has major themes running though it, and they are almost bound to appear in some form or another. The skill lies in detecting these themes and in being able to tailor them to the exam questions.

Choose the number of topics to study

Once you have identified the likely themes, you can plan which ones you need to study.

Prepare more topics than you are likely to use, so that you are not caught out by some of your expected topics not coming up. For example, if you have to answer three questions, it is a good idea to prepare at least five topics.

Organise your notes for each topic

It is impossible to revise properly when all your topics are mixed up in your notes and you have to jump from one source to another.

Prepare a separate folder for each topic. Even the act of preparing each folder and organising the information in it will help to consolidate the overall structure of the material in your mind. Include in each folder:

- your lecture notes
- notes from your readings
- photocopied material
- anything you may have written on the topic
- photocopies of past exam papers with the relevant questions highlighted.

Analyse the information in such a way that you can recall it

Many people seem to believe that information can enter them by some sort of osmotic process; that all you have to do is to have a lot of photocopied material and lecture notes and then go along and highlight the important points. This is a fallacy. It's very important that as well as identifying the important material:

- you analyse it for its key points
- **then write those key points down.**

This act of writing down the key points is the most powerful method of remembering information.

A swotting scheme might therefore be:

1. Read and re-read all your material on a topic. **Highlight the key points** with a highlighting pen. Make sure that they really are the key points and that you're not just homing in on words that look as though they might be important.
2. While you are highlighting the material, or immediately after working through a section**, make lists of the key points.** Make the lists succinct – include too much and you won't be able to recall it all. If the list is disordered or untidy, write it out neatly in a logical order. Use a numbering system, indented subheadings, colour schemes, coloured bubbles to surround items with – anything that will help you visualise it in your mind in the exam.
3. **Make up mnemonics to help you remember the items on a list**. Many people know readymade mnemonics , such as ROYGBIV (the colours of the rainbow) – fewer people make up their own. Yet they are a very powerful means of helping you remember information and moreover recall items in the right order when order is important.

   For example, here is a mnemonic for the electrochemical

series. It's a cunning one – it makes use not only of the initial letters of the elemental symbols but the second ones as well:

King Caractacus's nanny mugged Alfred's zany ferret, sneezing, probably having cured Helga's agnostic aunt.
(K, Ca, Na, Mg, Al, Zn, Fe, Sn, Pb, H, Cu, Hg, Ag, Au)

Mnemonics that you devise yourself don't need to be nearly as elaborate as this one. To make up a mnemonic:

- take the first letter of each item on the list of things that you want to remember
- make up a phrase in which the sequence of words uses those initial letters
- make your phrase as absurd as you can. It is much easier to remember something ridiculous.

4 When you've made your lists and mnemonics, **practise jotting them down from memory**. See how much of the original list you can recall. Look at the original when you need to, and then try jotting the whole list down again. With only a few repeats, you should soon be able to recall entire lists by thinking of their spatial distribution, their colour schemes and associated mnemonics.

5 **Use your recalled lists to help jog your memory of the key information**. If you can remember an item on your list but very little information about it, go back to your notes and fill in your knowledge in that area. Practise summarising information by using the one or two key words as pointers.

Sitting the exam

It is very common for people sitting exams, even the well-organised and comparatively calm ones, to make mistakes because of the stressful circumstances. There are a number of things you can do in the exam to maximise your effort.

1 **Make sure you understand the structure of the exam.** Take note of:

- the number of sections there are
- the number of questions you have to do from each section
- what the compulsory questions are.

2 **Read through the questions without panicking**. Be positive about them. Expect to be able to answer some of them; don't allow yourself to be panicked into thinking that you can't answer any of them. Look for the ones you can answer.

3 **Calculate the amount of time you have for each question**. In your calculations, allow for:

- The fact that some questions may be worth more than others, and therefore need proportionately more time.

- Ten minutes at the end of the exam to take up the slack and to allow you if possible to read through and correct what you've written.
- Slightly more time than calculated for each question to allow for the unexpected.

Note down the actual times-on-the-clock. Many successful exam-sitters make a note in their exam scripts of the actual time that they should expect to be finishing each question. It's easy to misjudge this under pressure and to run out of time.

4 **Choose the easiest question first**. This does a number of things:
   - This will probably be the question that you get the best marks for. Doing it first means you can get it behind you easily and without agony.
   - It gives you a sense of achievement while you're doing it, which has a calming effect.
   - If you leave it until the end (on the grounds that it's easy and you should have no problem with it) there is a good chance you'll run out of time and throw easy marks away.

5 **Analyse the question properly.** A huge number of marks are lost in exam essay questions as a result of not understanding the question properly. It's essential to **read and analyse the question properly.**
   - Read the question minutely to make sure you thoroughly understand the wording.
   - Look for the key words such as *analyse, compare and contrast, describe* (see Chapter 2 **Analysing an essay question**)
   - Remember that many exam questions include *two* keywords. The first nearly always requires you to give a description of something (*describe*, or *outline* etc.), the second keyword is nearly always *discuss*. Keep this in mind when you answer the question (see pages 14-15).
   - Keep in mind what you have to do when addressing issues using these keywords, and make sure that you stick rigidly to it. Don't lose marks by rushing into a question and misinterpreting it.

6 **Make an essay plan.**
   - This is **extremely important.** Before tackling any question, spend a few minutes jotting down an essay plan.
   - Don't take too long, because time is precious. Keep it simple; just make a list of a few keywords – perhaps in a simple concept map – to help you write a well-structured essay.
   - When writing the essay, stick strictly to your plan and avoid diversions. Many marks are always being lost by students who wander off the topic in the heat of the moment and lose their focus after a good start.

7 **Writing the essay.**
   - **Get directly to the point.** In your coursework essays you may

have spent time in thinking of original opening sentences. You have no time to do this in exams. In the Introduction rephrase the question and state the point of the essay: 'This essay...', 'In this essay...'

- **Focus on the main points.** Each one can be the lead sentence of a paragraph that then goes on to illustrate or amplify the point.
- **Use marker's language.** Use the terms and phrases that occur in the essay question.
- **In your Conclusion, summarise your argument using the phrasing of the question.** This shows your marker that you are on target.

8 **Keep to the times you have previously worked out.**
- Don't miss out any questions.
- If you are running out of time, the best strategy is to write two partial essays instead of concentrating on one and risk missing one out altogether. This is because the easiest marks to get for any question are those for the first half. Marks can often be gained for a series of only partially-developed main points.

9 **If you run out of time.** Jot down the main points of the rest of your essay. This shows the examiner the overall structure and the main content of your answer; it might gain you an extra mark or so.

Lecturers' expectations of exam essays

- Your essay will be clearly focussed on the topic and will deal with its central concerns as fully as possible within the limits of the exam time.
- Your essay will be the result of systematic revision of your course materials.
- Your essay will be logically structured. However, it is not expected that it will be as well-structured as a course-work essay because there is little time for redrafting.

Checklist: strategies for writing essays in exams

Preparing for the exam

- ✓ Identify possible topics.
- ✓ Choose the number of topics to study.
- ✓ Analyse the information so that you can recall it:
  - ✓ analyse it for the key points
  - ✓ write them down as lists
  - ✓ practise writing the lists from memory
  - ✓ make up mnemonics and practise them
  - ✓ check that the lists jog your memory.

Sitting the exam

- ✓ Establish the exam structure.
- ✓ Read through the questions without panicking.
- ✓ Calculate the amount of time for each question.
- ✓ Choose the easiest question first.
- ✓ Analyse the question properly.
- ✓ Make a quick essay plan.
- ✓ Get directly to the point.
- ✓ Focus on the main points.
- ✓ Keep to the times.
- ✓ If you run out of time:
  - ✓ do two part-essays instead of completing one
  - ✓ jot down the main points.
- ✓ Know what the markers' expectations are.

SECTION 2

Writing a report

Introduction: writing a report

Reports are the standard means of communicating in most institutions. Government organisations, hospitals, consulting firms, large corporations – all use the report format to transmit and obtain information. A report can be as brief as a single page or as long as a royal commission.

Always write with your readers' needs in mind, not your own. People read reports for one reason only – to obtain information as fast and conveniently as possible. We therefore have to structure and format reports with the various readers' needs always in the forefront of our minds.

This section of the book:

- describes how a technical report differs from an essay
- describes the basic skeleton of a report, and suggests strategies for deciding on an appropriate structure.
- gives the requirements for each of the sections of a report
- gives suggestions for planning and writing up a major report, such as a final year research project.

Other aspects of the writing process that are relevant to essay writing are covered in Section 3, **The Tools of Technical Writing.**
These are:

What is a report? How is it different from an essay?

6

CHAPTER

This chapter covers the differences between a report and an essay in their:

- appearance
- structure of the information
- purpose.

Most people are familiar with the essay format from school. But now, as a science or engineering undergraduate, you will probably be asked to write both reports and essays. Each of them can present the same sort of information, so what is the difference?

What is a report?

1. **It is a form of professional documentation.** In your future career you will often be required to write reports; you will rarely, if ever, have to write an actual essay. It is essential to be able to recognise the differences between them.
2. **It is a document designed so that your reader(s) can readily extract information from it**. This information could be:
   - an objective analysis or description
   - recommendations
   - your interpretation of the possible effects of something.
3. **In your professional life, you may have many different types of readers of your report.** They may:
   - need only an overview
   - need only some part of the information
   - have different needs and levels of understanding.
4. Therefore you have to:
   - **think of the readers all the time that you are writing**
   - **write from the point of view of their needs, not yours**
   - orientate them to what is in each part of your report
   - make it as easy as possible for each reader to extract the information that he or she needs.

5 This means that a report has to have:
- an informative Summary
- well-chosen subdivisions, with headings and subheadings
- the main points made obvious by indenting and boldfacing, together with listing within the text.

What is an essay?

1 **An essay is not designed as a working document.** The reader cannot readily obtain an overview or selected information from an essay. Many departments don't allow any headings or subheadings in an essay.

2 The reader has to read it in detail from beginning to end.

3 **It is more of a literary form of writing than a report.** You, the writer, without the aid of side-headings, have to rely solely on your wording to show the flow of your argument.

The grey area in the middle

Some departments may allow you to include a few side-headings in your essay. It can then start to look a bit like a report and the distinction between the two becomes blurred. But in general it can be said that a report is more heavily sub-divided and formatted than an essay.

Reports and essays: differences in structure and purpose

The Report	The Essay
A form of professional documentation.	Not designed as a working document. More of a literary form of writing than a report.
In professional life, often skimmed or selectively read.	Must be read in detail from beginning to end.
Heavily formatted: sections, subsections, text formatting. Designed to maximise the ease of extraction of the necessary information.	No visual formatting.

How to structure a report

CHAPTER 7

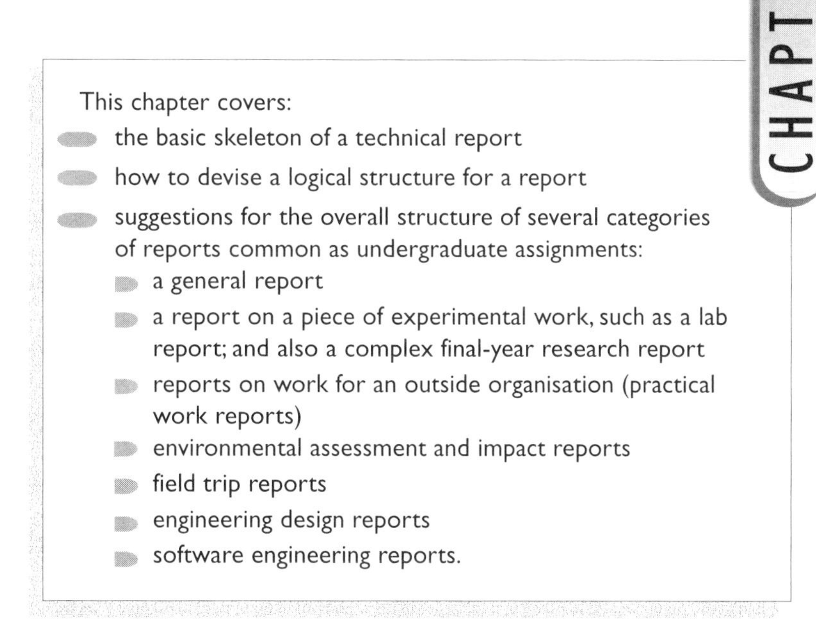

This chapter covers:

- the basic skeleton of a technical report
- how to devise a logical structure for a report
- suggestions for the overall structure of several categories of reports common as undergraduate assignments:
  - a general report
  - a report on a piece of experimental work, such as a lab report; and also a complex final-year research report
  - reports on work for an outside organisation (practical work reports)
  - environmental assessment and impact reports
  - field trip reports
  - engineering design reports
  - software engineering reports.

The basic skeleton

All technical reports have the same basic skeleton. Each one starts and ends with similar sections, but the headings for the middle part of a report will depend on its subject matter. This basic structure is used for all types of reports, from a simple first-year description of, for example, systems of waste-water treatment, to a complicated final-year report on an original research project.

The basic structure of a technical report

Title page
Summary (also called an Abstract)
Contents page
Glossary of terms (if needed)
Introduction or Background
Theory (if needed)
Middle sections (the headings of these sections depend on what type of report you're writing)
Conclusions
Recommendations (if needed)
Acknowledgements
References or Bibliography
Appendices

How to work out the structure of your report

Structure it from the point of view of your readers' needs. The objective of a report is to get information across to your readers. The structure of your report will have a large influence on this. Some reports need a very defined, rigid structure, for example, simple laboratory reports. Others require you to think up a suitable structure for the middle-sections; in this case you have to decide how you think your reader would best absorb the information. You have to ask yourself:

- **In what way is this story best told?**
- **What would be the most logical order for the reader?** Not the most obvious for you, but for the *reader*. Remember you could have arrived at this information in an order that may not be the best way to present it for a reader's understanding.

It is extremely important to remember that you are writing for your readers, not for yourself. Think of a sequence in which to present the information that is best for them, not for you.

Example: A first-year report on electric vehicles

This is a general report – the sort of assignment where you are required to report on an area of knowledge that you have read around, and possibly enquired about from outside sources.

Imagine you are required to write a report on electric vehicles. Your reading and enquiries have resulted in a lot of good, detailed information, but it is in a random order:

1. Will more power stations be needed?
2. Electric vehicles make less noise than those with internal combustion engines.
3. Present batteries have problems.
4. In California, 2% of each major auto-manufacturer's sales must be zero-emission vehicles in 1998, increasing to 5% in 2001 and 10% in 2003.
5. There are different sorts of batteries – lead-acid, nickel-cadmium, sodium-sulphur, and a number of others not yet fully developed.
6. Development of electric vehicles was boosted during the oil crisis in the 1970s.
7. Fuel cells are a possibility.

In this order, the information is difficult to absorb. In the same way, your reader will have difficulty in easily making sense of it if it is presented in an illogical order in your report.

How to structure information into a logical order

- **Always think of yourself as a story-teller.** A talking wolf lying in bed in disguise wouldn't be your first item if you were telling the story of Red Riding Hood. In the same way, you have to devise the most logical order for any story that you have to write up as a report.
- It **helps if you try to think in terms of *What Headings Should I Have*?** If you can think of what blocks of information you have, and then the order in which to put them, a good structure will come out of it.

Guidelines for structuring information

Step 1: Your first step may be to brainstorm your ideas for main sections onto paper, creating a concept map. At the same time, you can make a note of the possible subdivisions for each main heading.

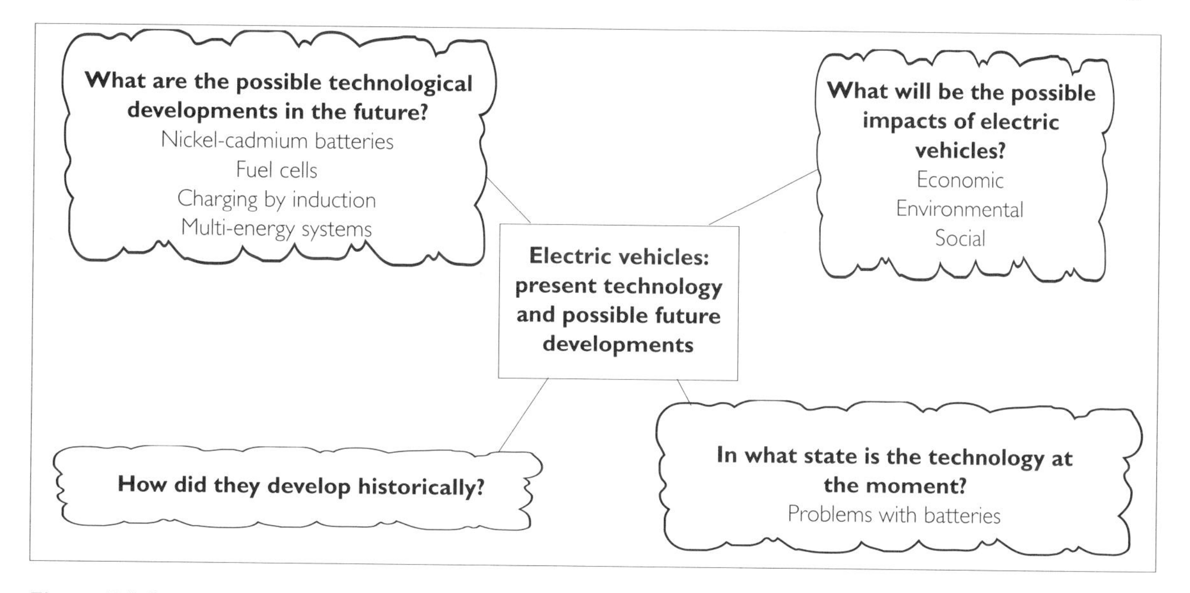

Figure 7.1 Concept map for the report on electric vehicles. The topic is written in the centre; the main aspects are placed at random around it, with some of their possible subheadings

Step 2: Next you need to get the sections into order, so that your story is logical.

Step 3: Some of these main sections will need subdivision, some will not. Now work out the headings for these subsections.

Step 4: (Figure 7.2) Now turn your concept map into a list of the main headings and subheadings. It should be a linear flow of ideas, and should look like a Contents Page of a book or report.

Step 5: In your finished report, the headings and subheadings should be numbered according to the convention for heading numbering (see Figure 7.2 and Page 53)

Example:

Electric vehicles: present technology and possible future developments

Thought processes	**Report structure**
An overview of the content of your report *(see page 51 for a sample Summary for this report)*	Summary
Think what you would put in the Introduction. *Obvious: Why are they being developed? The pollution caused by the internal combustion engine. Californian legislation*	**1.0** Introduction
Logical place to start: *The history of their development*	**2.0** The historical development of electric vehicles
This leads on to: *In what state is the technology at the moment? There are problems: examine them.*	**3.0** Current technology and its problems 3.1 Batteries 3.2 AC Motors 3.3 Other aspects
Which leads on to: *What are the possibilities for the future? How do we get over these problems? There are four aspects to look at.*	**4.0** Future technological development 4.1 Batteries 4.2 Fuel cells 4.3 Charging by induction 4.4 Hybrid vehicles
So far we've only looked at the technology. *What about the impacts on other things? Three aspects to look at.*	**5.0** Possible impacts of electric vehicles 5.1 Environmental 5.2 Economic 5.3 Social
What conclusions can I draw from my work? *(See page 62 for a sample Conclusions section for this report)*	**6.0** Conclusions
What reference material did I use?	**7.0** References
Who gave me help *(lent me books, gave me brochures, talked to me about it?)*	Acknowledgements
What other information do I have *that a reader might need, but which is so detailed it would clutter up the story?*	Appendices Appendix 1: Examples of recently developed electric vehicles Appendix 2: Drive system components

Figure 7.2 Final structure and thought processes for the report on electric vehicles

Different types of reports: structuring the middle sections

You may be asked to write several different types of report while you are an undergraduate. They will probably fall into a number of basic categories. Although the basic skeleton of all categories will be similar, you will need to devise structures for the middle sections of the reports that are appropriate to the type of report you are writing. The following section gives suggestions about how to do this. These are generalised suggestions; it is very important that you check your department's specific requirements.

The principle for structuring any report is the same in all cases:

- Do not deviate from the basic skeleton (see page 39) for the sections at the beginning and the end of the report.
- When you are devising appropriate headings for the middle sections remember:
  - **a** that you are telling a story (page 41)
  - **b** that you are telling that story from the viewpoint of what you think your reader needs, not from your own (page 40).

A general report

Here you are required to report on a specific area that you have read around, using the library and possibly contacting outside organisations for additional information.

Purpose of the report

To give a balanced, illustrated account of a particular area of knowledge.

Examples:

- solar power
- the potential for wave power in the United Kingdom
- electric and hybrid vehicles
- contaminated sites in New South Wales
- waste-water treatment systems.

These reports need to be structured like the electric vehicle example shown above – that is, for the middle sections you choose section headings that are appropriate to the material and the story that you are telling.

A report on a piece of experimental work

The experimental work may be:

- original research that you have carried out by yourself or in a group, typically in the final year of a degree
- a laboratory exercise.

Purpose of the report

- To describe your experimental work in sufficient detail for it to be repeated and verified by others.
- To draw conclusions from your data and findings.
- To place those conclusions in the context of related work in the area.

The sequence of sections in these types of reports is, typically:

- Summary
- Introduction
- Theory (if needed)
- Experimental procedure
- Results
- Discussion
- Conclusions
- References
- Appendices

A complex final-year research report

It may be more appropriate to structure your headings differently (see, for instance, the example Contents Page on page 54). But you still need to work within the framework of first describing your procedures, then presenting your results.

Separate experiments in the one report

- If your experimental work has a number of quite separate parts, group the Experimental Procedure and the Results sections for each experiment together.
- If appropriate, include a short Discussion section for each separate part and follow up with an overall main Discussion.

Good structure	Poor structure
Experiment 1	**Procedure**
Procedure	Experiment 1
Results	Experiment 2
Discussion of Expt. 1 (if wanted)	Experiment 3
	Experiment 4
Experiment 2	
Procedure	
Results	**Results**
Discussion of Expt. 2 (if wanted)	Experiment 1
etc.	Experiment 2
•	Experiment 3
•	Experiment 4
•	
Overall Discussion	**Discussion**

Combining the Results and Discussion sections
A more logical structure is sometimes achieved by writing a section called **Results and Discussion**. This allows you to present your results and immediately discuss them in context.

Reports on work for an outside organisation (practical work reports)

Purpose of the report
This is a report on work that you did for an outside organisation. The aim of this type of report is to describe the activities of the company and your work within it.

The sequence of sections could be:
- Summary
- Description of where you worked
  - **a** the type of enterprise and what it produces or does
  - **b** a brief description of the layout of the works and plant
  - **c** the staff organisation structure
  - **d** number of employees in the various sorts of work
- A full description of the work you did
- A full description of other work that you observed
- General comments on such things as:
  - **a** buildings
  - **b** layout of plant
  - **c** technical facilities
  - **d** amenities for staff (eating etc.)
  - **e** the state of industrial relations in the organisation.
- Conclusions (if needed)
- Appendices

Environmental reports

Purpose of the reports
To report on the condition, use and significance of a site or ecosystem.

There are two different forms of an environmental report:

1 An environmental impact report

Purpose of the report
To assess the impact of a proposed development or change in use.

2 An environmental assessment report

Purpose of the report

- To provide basic information on a natural environment for the planning department of the local authority. No development is planned for the area at present; it is simply seen to be undervalued in terms of conservation or public interest. The site may be in private or public ownership.
- To give an objective and subjective assessment of the 'value' of the site or ecosystem.
- To encourage protection of the area.

Suggested sections for these reports would be:

- Summary
- Introduction
- Description of the area – its locality, geography and present condition.
- Assessment of value – as an amenity, natural system, aesthetic feature etc., requiring comparison with others nearby.
- Analysis of its uniqueness – providing some scale of its importance.
- Analysis of its significance, urgency of action, irreversibility of action and extent of loss to society or community if not protected.
- For an *Impact Report*: An account of the impact of the development, and the environmental safeguards to be adopted.
- Recommendations will probably be needed.
- Conclusions
- References
- Appendices

A field trip report

Purpose of the report

To report on a field investigation. In the biological sciences they generally deal with ecological phenomena, and in the earth sciences, with aspects of geological formations and fossil life.

It is essential that these reports should not be a regurgitation of general material, but should present your personal observations.

- **Biological sciences.** Reports are usually required in the standard format of a description of a piece of experimental work (see page 43-44).

- **Earth sciences.** Suggested headings would be: Summary, Introduction, Landforms, Lithology and Faunas, Structure, Discussion, References, Appendices.

Design reports: engineering/software development

Purpose of the report

To give the specifications for an engineering design, or for software development.

To design is to create the description of something that does not yet exist. Therefore, these reports are quite different from other types. They are:

- open-ended
- require creativity and individual judgement
- must satisfy a particular set of rules.

Remember that a design report has a clear purpose: subsequently another team or person will expect to be able to construct what you have designed *without having to refer back to you*. This is how your marker will judge your report – as though he or she is a client. It is essential to transmit your ideas clearly.

A general format for engineering design reports cannot be given, since it will depend on the individual departments and lecturers. Some departments regard your workbook as the design report; others require an assembling of your material into something that more closely resembles the other types of report. It is essential to conform with what your department needs.

General points

- **Your calculations or coding must be presented in such a way that someone can follow them.** You have to anticipate the way in which your marker is most clearly going to understand your reasoning, and then present the sequence of your design.
- Drawings must be self-explanatory.
- You need to quote and fully reference formulae, standards or codes (see page 110 for guidelines on referencing codes and standards), plus any assumptions you may have made.

Checklist: how to structure a report

Work out the headings for the middle sections of your report:

✓ remember you are telling a story
✓ write it from the point of view of your readers' requirements, not yours
✓ brainstorm ideas for headings and subheadings onto paper
✓ put them into order
✓ make them into a list that looks like a contents page.

Know what is required for the following types of report:

✓ a general report
✓ a report on experimental work
✓ a report on work for an outside organisation
✓ an environmental assessment or impact report
✓ a field trip report
✓ a design report.

Requirements for each section of a report

CHAPTER 8

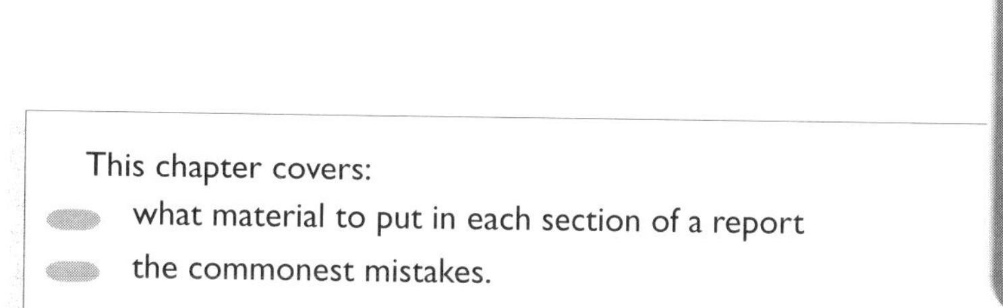

This chapter covers:

- what material to put in each section of a report
- the commonest mistakes.

This chapter describes what you need to include in each of the sections of your report. Use the previous chapter to understand the basic skeleton of a report's structure.

Title page

The Title page gives:

- the title of the report
- the author's name, department and university
- the course or paper number
- the date
- if needed, a declaration that it is your own work.

These should all be logically and clearly arranged on the page, as shown in the example.

Guidelines for the Title page

1 **A good, informative title is essential.** It orientates your reader immediately to what your report is about. A poor title can be misleading if it is unspecific.

Poor, unspecific title	Informative title
Flow meter experiment	Measurement of the coefficient of discharge in a Venturi meter.
Waste-water treatment	The Newtown waste-water treatment station: treatment processes, output standards and environmental impact.
Wave power	Wave power: its potential in The United Kingdom.

2 Don't put the title in quotation marks.

3 **Signed declaration**. Some universities and departments require a signed declaration that the report is the student's own work. Put it at the bottom of the title page.

Suggested wording:
I declare that this report is my own unaided work and was not copied from or written in collaboration with any other person.

Signed............................

Example of how to lay out a Title page

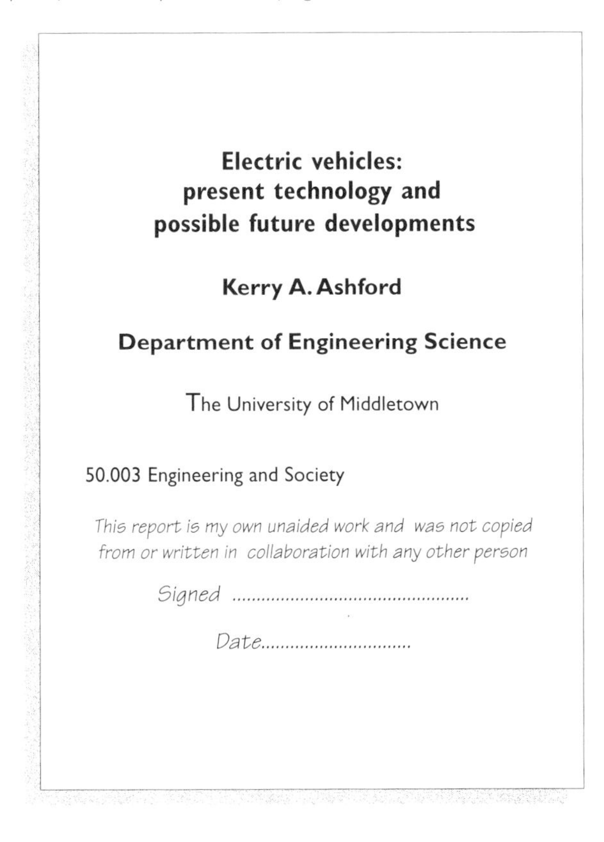

Electric vehicles:
present technology and
possible future developments

Kerry A. Ashford

Department of Engineering Science

The University of Middletown

50.003 Engineering and Society

This report is my own unaided work and was not copied from or written in collaboration with any other person

Signed ....

Date..............................

Summary *or* Abstract

These two words are often used to mean the same thing. This book uses the word Summary.

The Summary is an extremely important section because it gives a brief overview of the substance of your report. It is a very powerful aid to help the reader's understanding of the whole report. **Summarising skills are essential to good report writing.**

What makes a good Summary?

1 **A good Summary is a self-contained synopsis of the substance of the report**. Your reader should know from it:
   - the main features of the investigation
   - what were the results or the outcome of it
   - your main conclusions.

2 **It should be concise and brief (100-200 words).**

Example: a good Summary

The following example shows a summary of a short report, the report given as an example of structuring in Chapter 7 **How to structure a report** – a first year report on electric vehicles.

A one-sentence statement of what the report is about.

Now follow it up with a logical progression through the main aspects of your report.

Summary

This report examines electric vehicle technology, possible future technological developments, and the environmental, economic and social impacts.

No current electric vehicle can equal the performance of an internal combustion engine. Lead-acid batteries are heavy and thus reduce the payload, have a low range at high acceleration, low charging speeds, and a comparatively short life. Development work is taking place in different types of battery, AC motors, hybrid vehicle technology, fuel cells and charging by induction.

A country adopting electric vehicle technology will need a comprehensive network of recharging points, and probably increased generating capacity. Vehicle emissions will be reduced, but not necessarily overall pollution, since more power stations may be needed. However, there may be substantial benefits in those countries where hydropower is a main source of energy. Social attitudes are expected to move away from high-performance cars towards zero-emission vehicles. It is concluded that electric vehicle technology will be competitive by early next century.

Now finish off with the main conclusion.

Note: do not refer to any figures or cite any references in a Summary

What are the characteristics of a poor Summary?

A poor Summary just tells the reader the structure of the report – rather like a series of chapter headings - instead of giving an overview of the real substance of it. It is descriptive rather than informative.

Example: a poor Summary

This summary gives no real information. It describes only the structure of the report, not its findings. Beware of phrases like such-and-such is described, is analysed etc.

Summary

This report examines electric vehicles. The current technology is examined, and the problems are described. There is future potential in new types of batteries and other areas are also analysed. The environmental, social and economic impacts are considered.

General guidelines for writing a Summary

- First state what the report is about. If you are a beginner at writing reports, it is a good idea to form a logical structure by starting off with 'This report describes.....' or 'In this report.....'
- Next give the main information, probably in the order in which it is presented in the report. Make sure you give real information made up of hard facts.

- Then end by giving the main conclusion(s). Leave the less important ones for the Conclusions section.
- Do not refer to illustrations or references in the Summary.

To write a Summary of an experimental investigation say *in this order*:

- What the report is about, e.g. This report describes the modelling of a car suspension unit.
- How you did it (a brief description of the method).
- The results you got. Make sure you include the main quantitative values, not just vague generalisations. This is very important.
- The main conclusions (for instance whether your results are consistent with current theory).

Where to place it

The Summary is usually placed on a separate page immediately after the title page.

Common mistakes: the Summary

- Making the Summary too vague (see the examples of a good and a poor Summary, page 51).
- Not giving quantitative values for the results.
- Mentioning specific figures, as in 'Figure 3 shows that...'. You should not refer to specific illustrations nor cite references in the summary.

Acknowledgements

If you are writing a major report and wish to acknowledge your supervisor, other members of staff, people in industry and other organisations, etc., it is gracious to place the Acknowledgements section on a separate page immediately after the Summary. If it is a short report, and you have fewer acknowledgements, you can place the section immediately after the main text and before the References section.

This is a formal report; don't be too colloquial or fatuous in your thanks, and always include surnames.

Contents

This should clearly set out the sections and subsections of your report, and their corresponding page numbers.

Guidelines for setting out the Contents page

- **Place the page numbers at the right hand side of the page.**
- **Number the sections by the decimal point numbering system.**

The main sections of the report are given Arabic numerals. The sub-sections are denoted by putting a decimal point after the section number and another Arabic numeral:

1.0 Title of first main section
  1.1 First sub-heading
  1.2 Second sub-heading

2.0 Title of second main section
  2.1 First sub-heading
  2.2 Second sub-heading
    2.2.1 First division in the second sub-heading
    2.2.2 Second division in the second sub-heading
    2.2.3 Third division in the second sub-heading
  2.3 Third sub-heading

3.0 Title of third main section

Don't go beyond a three-number / two decimal point (e.g. 2.1.1) sub-sub-heading. It gets too complicated. If you find that your text needs more subdivision than this, use bold side-headings – perhaps indent them and the text – but do not list these headings on the Contents page.

- **Conventions for numbering the pages.**
  a All the preliminary pages, i.e. those preceding the Introduction (Title page, Summary, Acknowledgements, Contents page, List of illustrations etc.) are numbered in lower-case Roman numerals (i, ii, iii, iv, v, etc.). The first page that is counted is the Title page, but do not label it as such at the bottom of the page; leave it blank.
  b Number all the remaining pages of your report with Arabic numerals, making page 1 the first page of the Introduction.
- **Appendices** (Note the use of the words: it's one Appendix, two or more Appendices). Don't just call them Appendix 1, Appendix 2 etc. Each should have a title describing the contents of the appendix, e.g. **Appendix 1: Sample Calculations**.
- **Figures** Don't list individual figures on the contents page. If you have a lot of figures and want to include the page numbers of the them, have a section called **List of Illustrations** (see page 55 for Guidelines); mention it and give its page number on the contents page.

Example: Contents page

This is the Contents page from a complicated final year research project. It needs subdivision down to sub-sub-heading level. It is a good model for both simple and complex reports, because it shows all the conventions for a Contents page.

Summary and Acknowledgements not listed. Place them on separate pages before the Contents page

Roman numerals for everything before the Introduction

Start numbering sections and pages at the Introduction

Decimal system used for main headings, sub-headings and sub-sub-headings

Indenting of the left margin shows the hierarchy of each section's sub-headings

No section number for appendices

Each appendix numbered and titled

Contents

Common mistakes: the Contents page

- Page numbers left out.
- Irregular indenting of section headings and subheadings.
- Incorrect use of the decimal system for numbering headings
- Untitled appendices – listed only as Appendix 1, Appendix 2 etc.

List of Illustrations

The term *illustrations* includes both tables and figures. This section may be needed if your report has many graphs, tables, line drawings etc. Chapter 12 **Illustrations** has guidelines for deciding on and producing the different types, and referring to them in the text.

Guidelines for preparing a List of Illustrations

- Title it *List of Illustrations* if it contains both tables and figures.
- If it contains only figures (graphs, photographs, line drawings, maps) call it *List of Figures*; if only tables call it *List of Tables*.
- If it contains both figures and tables, list all the figures first, and then list all the tables.
- List the number, title and page of each illustration.
- Place the List of Illustrations immediately after the Contents page. If both of them are brief, put them on the same page, with the Contents first.

Glossary of Terms

(The Glossary of Terms can also be called **List of Symbols** if it's in a mathematically-based report).

In this section, define any terms that you use in the main body of the report. These could include:

- Specific technical terms that you think need to be defined.
- Greek or other symbols
- Abbreviations. These are words formed from the initial letters or parts of other words, e.g. PCR: polymerase chain reaction; SEM: scanning electron microscope. If you are going to use the abbreviation throughout the text, it must be defined at the first point in the text that the term is used. Put the abbreviation in brackets after the term. *Example: The sample was examined in the scanning electron microscope (SEM).* Thereafter it is acceptable to use only the abbreviation. But if you use a number of them, it helps the reader if you list them in the glossary as well.

Introduction

This needn't be long. It describes the background to your work, starting with the broad context of your investigation and leading up to the hypothesis.

Guidelines for writing an Introduction

- Say what the report is about.
- Start by putting the reader in the picture, perhaps by giving a brief historical overview of your area.
- Choose relevant facts from the literature and assemble them into a logical sequence. Cite the sources for each statement. If you are writing a laboratory report, avoid using your manual as the sole source.
- Introduce the methods, but only as an overview. The details of the method belong in the Material and Methods/Procedure section.
- Introduce the hypothesis.
- State the scientific objective of your work. Never write that the objective is 'to measure this or that' or 'to learn about the technique'.
- Give an outline of what is in the report (e.g., Section 4 describes...).

Theory (if needed)

If you are writing a technical report you will probably need to explain some of the theory behind your investigation.

Guidelines for writing the Theory section

- **Don't include too much.** Imagine you are writing for a busy professional engineer or scientist who is not familiar with the intricacies of your subject, but who has a good technical knowledge. Include as much theory as you think such a person will need.
- **Avoid uncritical word-for-word copying** of large amounts from textbooks.
- **If your report needs only a small amount of theory**, it may be included in the Introduction.

Middle sections of the text (this should not be used as a heading in reports)

You have to decide what are appropriate section headings for this part of your report. They should present the material in the order in which the reader will best understand it and form a logical progression of concepts. Chapter 7 gives suggestions about how to do this.

A report on experimental work has a classical sequence of section headings:

- Materials and Methods (or Experimental Procedure)
- Results
- Discussion.

Note: combining the Results section with the Discussion *may sometimes make your report easier to understand. Call it* ***Results and Discussion.***

The requirements for each of these sections are given below.

Materials and Methods or Experimental Procedure

Describe exactly how you carried out the investigation – the materials and equipment, and what you did with them. It is one of the easiest sections to write.

Guidelines for writing the Materials and Methods section

- **Include enough detail of the equipment and the procedure so that another person can repeat exactly what you did, working from only your description.** This is the only way to verify scientific and technological investigations. A comprehensive Procedure section is therefore one of the main ways to give your report credibility.
- **Write it in the narrative form**, not as a series of instructions such as you would find in an instruction manual. (For an explanation of this, see the Common mistakes box below).
- **Describe the equipment.** (Note, however, that in a large report, technical details of standard equipment are often placed in an appendix.) Don't include trivial details (see below, the Common mistakes section).
- **Describe the material or samples or specimens that you used.** Give all the relevant details, e.g. composition, specification, descriptive anatomy, etc.
- **Describe how you obtained the data.**
- **Describe how you treated the data.**

Common mistakes: Experimental Procedure

- Not including enough detail.
- Giving unnecessary, trivial detail.

Examples

The computer was switched on by using the switch at the back, and then the model was chosen by clicking on the icon with a mouse. (Self-evident procedures)
Clean test-tubes were carefully filled with reagent *x*. (The concepts of *clean* and *careful* can be taken as read.)

- Writing it as a series of instructions. Laboratory manuals and other types of instructions for an investigation are often written as a series of instructions.

Example of Experimental Procedure as written in the lab manual

1. Measure width, thickness and length of specimen.
2. Pass specimen through hand-rolls with (a) width parallel to rolling direction and (b) one end aligned with edge of the rolls at the widest opening end.
3. Measure thickness profile.
4. Place specimen in 600 °C furnace, 15 min; water quench.
5. While heating specimens, prepare Graphs 1a, 1b and 2.
6. Etch in Tucker's reagent.
7. Mark specimen at 1 cm intervals along its length.
8. Record: No. of grains/linear cm across the specimen's width at each centimetre.

Many students copy these instructions word-for-word from the manual. The important point to remember is that **you must write it, in the past tense, to describe what you actually did**.

Example rewritten in a style suitable for a Procedure section:

Experimental Procedure

The width, thickness and length of the specimen was measured to the nearest *x* mm. It was then passed through hand rolls with its width parallel to the rolling direction and one end aligned with the edge of the rolls at the widest opening end. The thickness profile was then measured to the nearest *x* mm. After heating for *x* minutes at 600 °C and water quenching, the sample was etched in Tucker's reagent for *x* minutes. The number of grains per linear centimetre was then recorded at centimetre intervals along the sample's length.

Notes:

- It is written as a narrative, not as instructions.
- The abbreviations used in the lab manual (e.g., min., No.) are written out in full.
- Instruction Number 5 in the lab manual was not part of the experiment. It is therefore not included in the Experimental Procedure section.
- The manual gave no instructions about the accuracy to which you were required to measure. You had to decide. This information must be included in your Experimental Procedure section.

Results

- In the Results section you present the information that is significant, i.e. the information that leads to conclusions about your investigation.
- The data you originally recorded – the raw data – often have to be converted by calculation or statistical treatment to a more useful form for your report. Do not include large amounts of raw data in the Results section: put it, together with a summarised version of the treatment, in an appendix.

Guidelines for writing the Results section

- **Draw the graphs and tables** you need from the experimental data you recorded. Chapter 12 **Illustrations** gives details about this.

- **Write the text.** Many people make the mistake of having a Results section made up of only graphs and tables. Graphs and tables merely present data; they don't state results. They **must** be linked by explanatory text. **This is important.** Moreover, for your text, don't just write 'The data are given in Figure 1 and Table 1'. Here are guidelines for writing the text in the Results section:
  a **State the results briefly.** Don't describe the curves themselves (as in, 'The curve showed an initial increase, followed by a steep decline'). State specifics: 'The cell density showed an initial increase, followed ...'. Don't repeat the data in the tables.
  b Write something about each figure and table.
  c **Refer to each one in the text by its figure or table number** (Chapter 12 **Illustrations** describes how to do this).

- **Do not discuss the results, just present them.** The place for comparing your data with theory and for interpreting them is in the next section, the Discussion.

Common mistakes: Results

- Presenting only graphs and tables, without any linking explanatory text (see above).
- Putting the raw data in the Results section and the interpreted data (graphs etc.) in an Appendix. Put graphs in the Results section and complex raw data in an Appendix.

Discussion

In the Discussion you say what you think your investigation means. This is where you comment on your results and interpret them in relation to the objectives of your work. It is an expression of your ideas, and is probably the most difficult section to write.

Guidelines for writing a Discussion

- Compare your results with values from references, or with what you might expect. Then discuss the differences (see the box for what to do when you get 'wrong' results).

What to do if an experiment or investigation has 'gone wrong'

It's very common to worry about how to present your work if you don't get the results you expected. Don't be concerned; you can still write a good report. Write it up as follows:

- In the Results section of your report, just describe your results exactly as they were. Don't mention in this section that they were different from what was expected.
- In the Discussion section, state that they were not as expected and give your reasons why you think they were different.
- State in the Conclusions section that the results were different from expectations, and briefly give your reasons as to why you think this was so.
- Do not feel uneasy or blame yourself or others for what went wrong. Simply consider the sources of error and the extent to which they affect the data. Some potential sources of error are:
  - **a** Experimental errors such as lack of precision in adding reagents.
  - **b** Sampling errors – with fewer replicates you are more likely to miss the real mean value.
  - **c** Errors of measurement such as difficulty in reading the exact figure on an instrument.
  - **d** Errors in recording – entering the wrong number on your records.
  - **e** Errors in computation – doing the calculations wrong.
  - **f** Differences in procedure – where you may have done something differently from the procedure in the manual.

- **Shape of the graphs.** Comment on the form of the graphs and the pattern of the results. Do they match your expectations? Do they show unexpected variations?
- **Error.** What are the significant sources of error in the results? (See box above)
- **Error analyses.** If you have used an error analysis, this is the place to cite it as a measure of the significance of the results. If you have to compare results, it is very important to use the magnitude of the errors in your comparison (see page 92 for graphical representation of error measurements).
- Does your work agree or disagree with other work in the field?

- What are the effects of any assumptions or approximations you may have made?
- What are the experimental limitations?
- Can you suggest any changes in the procedure that would give better results?

Conclusions

This is as important a section as the Summary. Most professional people who are too busy to read a whole report will read the Summary and probably the Conclusions. From these they will expect to understand the fundamentals of your work.

Guidelines for writing a Conclusions section

- Make sure that you call it **Conclusions** and not **Conclusion**. The latter might lead you into writing a paragraph in the manner of the conclusion to an essay.
- This section is made up of a series of the conclusions that you have drawn in the Discussion section of your report. Each one must be drawn directly and logically from your findings, and be as precise and as quantitative as possible.
- There mustn't be any waffle. This is important. Common ways of waffling are writing things that are not direct conclusions, such as:

  This was an interesting subject for a report and it taught me a lot.

  This was an easy experiment and the only way mistakes could have been made was in calculations by the student.

- It is best written as a list that is either bulleted or numbered. Some departments or staff members won't want it in the form of a numbered list. If this is so, it is important to remember that it should not be like an essay conclusion. You still need to write it in the form of a structured series of direct conclusions, one per paragraph.
- Start the list with the most important, unequivocal conclusions that you can draw from your work.
- Work down to the least important and more tenuous conclusions.
- Make each point as succinct as possible.
- Don't introduce any new material.
- All the conclusions that you list here will have already been pointed out and discussed somewhere else in the report. Don't worry that you are repeating material when writing this section; it's inevitable. Remember that this is another sort of summary.

Example

The following example shows a suggested Conclusions section for the first-year report on electric vehicles given as an example of structuring in Chapter 7 **How to structure a report**.

Conclusions

- Electric vehicle technology will become competitive by early next century.
- It is a very necessary development for the reduction of vehicle emissions.
- New developments in battery technology, fuel cells and methods of increasing the speed of recharging should improve performance.
- A network of recharging points will be needed, and probably more generating capacity.
- Social attitudes are expected to move away from high-performance cars towards zero-emission vehicles.

Common mistakes: Conclusions

- Treating this section as something called Conclusion (without the *s*), and writing generalised material, e.g.

  This experiment demonstrated the importance of this principle in biology.

- Writing vague statements that should be supported by detail, e.g.

  The strength of Sample 1 was much greater than that of Sample 2.

- Including new material, particularly the interpretation of results. Such material should be in the Discussion.

Recommendations

You may be asked to give recommendations for future work or planning. Present this section as a bulleted or numbered list, with each recommendation as a concise statement.

References or Bibliography

Here you list the sources of your information.

Part of the culture of science is that the conventions for laying out these lists are strictly observed by many lecturers, and you can lose a lot of marks if they're set out in the wrong way. Chapter 13 **References** explains how to do it.

Appendices

The appendices are for detailed material that would otherwise clog up the story told in the main body of the report. For that reason, they are isolated right at the end of the report.

The important points to remember

1. A reader should be able to read a report without ever having to look at the appendices. They are there only for the specialist reader who may need detailed information. Readers should not be made to flip between the main text and the Appendices to find the information necessary for understanding the main text (see Common mistakes box).

2. Each appendix must be referred to at the relevant point in the text.

3. The type of material that is put in appendices includes:
   - raw data and a summary of their treatment
   - lengthy tables
   - detailed descriptions of equipment
   - mathematical derivations
   - detailed error analyses
   - sample calculations
   - computer program listings
   - detailed maps and charts.

Common mistakes: Appendices

- **A very common fault:** writing in the report something like 'The data obtained in this section of the investigation are given in Appendix 3'. The reader then has to go to the end of the report and find the relevant appendix and then – even worse – has to plough through and interpret a mass of detailed information before coming back to the story in the main section again and trying to pick up the thread.

  This feature – forcing the reader off to an appendix to extract and interpret information – is something that is very familiar and maddening to many professional scientists and engineers. It is essential to remember that the general reader should be able to understand the main body of the text without having to refer to the appendices. They exist only for the specialist reader needing detailed information.

  What to do:

  Summarise the contents of an appendix at the relevant point in the main body of the report. Write something like 'The data obtained are summarised below. The detailed data are given in Appendix 3'.

- Not referring to an Appendix in the body of the report. Appendices shouldn't be left dangling and forgotten;

each one must be referred to in the text at the appropriate point.
- Putting too much in the Appendices. Material should be there for a reason. The Appendices are not a sort of rubbish bin into which you throw material you don't know what else to do with.

Checklist: requirements for each section of a report

Each section of a report has specific requirements for its content.

Use this chapter for the requirements and the commonest mistakes in:

✓ the Title page
✓ Summary *or* Abstract
✓ Contents page
✓ Glossary of Terms
✓ List of Illustrations
✓ Introduction
✓ Theory
✓ Materials and Methods *or* Experimental Procedure
✓ Results
✓ Discussion
✓ Conclusions
✓ Recommendations
✓ Acknowledgements (for References and Bibliography *see* Chapter 13 **References**)
✓ Appendices.

Planning and writing up a major report

CHAPTER 9

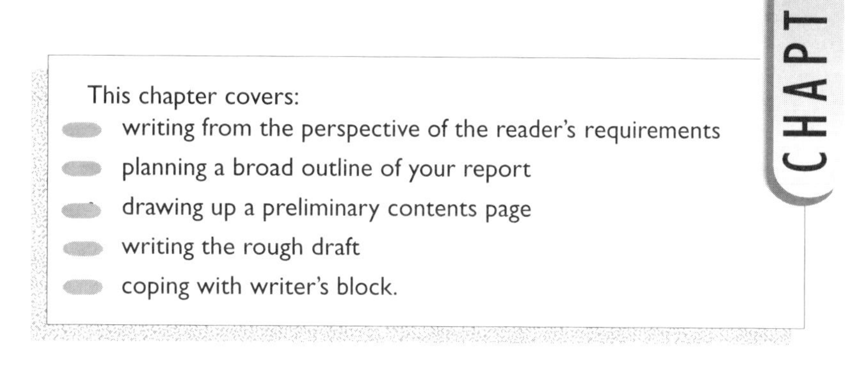

This chapter covers:

- writing from the perspective of the reader's requirements
- planning a broad outline of your report
- drawing up a preliminary contents page
- writing the rough draft
- coping with writer's block.

The deadline for your report is approaching, and you have a mass of material. Many students get fazed at this point and don't know how to bring it all together. There are a number of useful strategies for coping with this phase of your work.

1 First of all, stop working on your project

Many students don't want to stop. There are two reasons:

- Just a bit more work will tie it all up better. **Remedy:** Accept that research work opens more areas than it closes, and at some time you have to call a halt.
- An innate fear of the writing process. **Remedy:** Just attack it. It's the mechanical process of sitting down and attacking it that makes the words come and the ideas flow, and pushes the process forward.

2 Write in tandem

If you can't stop the practical work completely, at least start writing in tandem with it.

3 Remember how much time it will take

Don't put off the writing process to the point where you're rushed at the end. Allow for far more time than you think you'll ever need. Remember that producing a document, particularly the fiddly final stages, always takes far, *far* longer than anyone ever imagines.

Steps in the organising process

1 Sit and think

The amount of writing that you will eventually do will be inversely proportional to the amount of planning. Whatever you do, don't just start writing text. The result will probably be a mess and will need extensive rewriting.

Remember to put yourself in the readers' place. You are not writing to please yourself; you have to think always about your **readers**. Ask yourself:

- What are they most going to want to know about your project?
- What questions would they want answered?
- What order of sections would make the most logical story for them?

2 For whom are you writing?

It is usual in many departments to have two markers of a major report – a staff member who is expert in your field, and one who is not. Write for the second marker. Even if you do not have one, write as though you do. In other words, write for a busy professional person who has a good, broad knowledge of your discipline, but no special expertise in your particular area.

This means that:

- You have to set the scene for this person. It is a common mistake to make a report unintelligible by dealing only with unlinked detailed material.
- You have to decide how much background is needed to enable him or her to understand what you have done.
- BUT: she or he is too busy and too professional to want or need a simplistic explanation of the background to it.

3 Plan a broad outline

Use Chapter 7 **How to structure a report** for this step. The processes explained in it should be scaled up when preparing a large report.

4 Draw up a preliminary contents page

Many students find that the most useful next stage is to draw up a preliminary Contents page.

- This is merely a formalisation of your structured list into a numbered list of section headings and sub-divisions (see page 54 for a sample Contents page, and page 53 for the conventions for numbering the sections).
- If you are writing up on a word-processor, think about using the Outline mode that all major wordprocessing software packages have. It's a very useful tool if you're writing up a fairly long report because it forces you to think logically. While you construct your list of headings and subheadings, it will:
  - **a** Automatically assign numbers to each section in the standard decimal system of numbering for reports (see page 53). You can customise this.

b Rearrange the numbers if you change the position of any item in the list.

c Allow you to construct your whole document by filling out under the headings.

d Construct a finished contents page with page numbers when you've finished writing up.

5 Starting to write: the first draft

You now have your plan – where do you start on the actual writing? Here are guidelines for getting to grips with the writing process.

a Don't write the sections in the order in which they appear in the report

Many people think that they ought to begin by writing the Introduction. Don't try it yet; it's one of the more difficult sections to write, because you have to include reference to other people's work.

b Start with the easiest section

This is a very positive way to start the writing process. It will give you a sense of achievement while you do it, because you won't be so prone to the writer's block that afflicts everyone – even experienced writers – from time to time. Sort out your records for this section and focus your thoughts onto this one and no other.

Which is the easiest section?

- **Almost invariably the easiest will be the section describing your method of investigation.**
- Next write the **Results** section.
- Another relatively easy section to write up is the **Theory** section.
- **You are now beginning to feel that you're making some progress.** This helps to make you feel positive about tackling the difficult sections – the Discussion and the Introduction. In fact, the writing you've done so far has probably helped you consolidate your ideas for these sections.
- **Leave the Summary until last**.

c Treat each section as a story in miniature

- **Write each section in the way that your 'second marker' can best understand it (see Point 2 on page 66).**
- Make a list of the main points to be covered in the section. Use each of these points as the focus for each of your main subsections or paragraphs.
- Start each section by describing what you are going to cover

in that section. This gives an overview and makes it much easier to understand the material.

d Structuring the text within a section

A report should be written so that the reader can readily access the information in it. In addition to the structuring by sections and sub-sections, a powerful technique is to structure within the text by using boldface type and bullet (dot) points.

Material written as solid, 'black' text	The same material formatted for ease of access to the information
Negative environmental aspects of wind power While wind power has many advantages (in that it is a form of sustainable energy production), there are a few negative environmental effects related to its use. Windmills have a significant visual impact, especially as a wind farm may contain anything from fifty to hundreds of windmills clustered together (they are around 30 m high). Careful selection of the sites for wind farms can reduce this effect. The blades of a windmill can generate noise. This has been reduced by improved design, and most of the sound is masked by wind noise. This is generally not a problem from about 300 m away. TV and radio interference can be caused by metallic blades. However, most modern windmill blades are now made from non-metallic composites which virtually eliminates this problem. Bird strikes can, in some sites, be a significant issue, birds of prey being particularly affected.	**Negative environmental aspects of wind power** While wind power has many advantages (in that it is a form of sustainable energy production), there are a few negative environmental effects related to its use: • **Windmills have a significant visual impact**, especially as a wind farm may contain anything from fifty to hundreds of windmills clustered together (they are around 30 m high). Careful selection of the sites for wind farms can reduce this effect. • **The blade of a windmill can generate noise.** This has been reduced by improved design, and most of the sound is masked by wind noise. This is generally not a problem from about 300 m away. • **TV and radio interference** can be caused by metallic blades. However, most modern windmill blades are now made from non-metallic composites which virtually eliminates this problem. • **Bird strikes** can, in some sites, be a significant issue, birds of prey being particularly affected.

Points to note about the use of bullet points:

- A report written entirely in the style of the left-hand column is visually daunting.
- The material in the right-hand column has needed no alteration in the wording from the 'black' text.
- This technique, although powerful, needs to be used intelligently. Unintelligent bullet-pointing and bold-facing can fragment a text into chaos.

e Make a note at each relevant point in the text of each of the references you are going to cite

At this point it doesn't matter whether you are required to use the author/date (Harvard) system or the sequential numbering system for citing your references (see Chapter 13 **References**, for a full explanation of referencing). The details of the method can be sorted out in the rewriting stage (see Chapter 14 **Revising the first draft**). In these early stages it's important only that you make a note of which ones you are going to use and where they are going to be placed.

f Plan and make a rough drawing of each illustration

Chapter 12 **Illustrations**, explains how to:

- decide whether to use graphs, tables or text
- decide on the type of graph you need
- number, title and label your illustrations
- refer to your illustrations in the text.

The details can be sorted out later. At this early stage it is important only to plan the types of illustrations and decide where they are going to be placed.

Coping with the feeling of not knowing how to start writing

This is a very common feeling at the beginning of the writing process. There are two main strategies for coping with it:

1 **Write the easiest section first** (see Step 5, page 67). Feeling blocked can often be a reaction to the enormity of the task. Breaking it down into manageable chunks is a major help. Once you have written something, continuing the process is much easier than overcoming the first hurdle.

2 **Discuss your work with someone else.** Very many people find that they can explain their work to another person, but can't write it down. If you find that you are like this, put it to positive use. Actively go and find people to discuss your work with; this often helps you to clarify the structure of something and get the ideas to flow. Then write soon afterwards.

Coping with procrastination

It seems to be an integral part of the writing process that even experienced writers find excuses not to start writing just yet: the kitchen needs tidying, the area around the keyboard needs clearing, another cup of coffee is in order. Accept that this is nothing unusual and that you are going to have to discipline yourself to ignore these less than pressing reasons and just sit down and start.

Coping with writer's block

Don't feel unique and miserable when you arrive at a place in your writing where you come to a halt. Even experienced writers suffer from this sense of suddenly being stuck. There are two strategies that may help.

1 If you come to a halt in the middle of a passage, make some notes of what you want to say and carry on to the next bit, which may flow more easily. When you come back to it later, it often clicks easily into place.

2 Most important of all, don't struggle on feeling inadequate. If you think that, like a number of other people, you have a problem, talk to a member of staff well before the submission deadline. Many things can contribute to writer's block, and you can be helped.

Using a wordprocessor

There are several pitfalls to be avoided when using a wordprocessor:

1 **Save regularly and keep backups.** Far too many students realise much too late the pitfalls of not saving and backing up often enough.

2 **If you are using a departmental computer,** bear in mind that as the deadline approaches, the demand will be very heavy indeed.

3 If you are using both a departmental computer and one at home:
   - Make sure that the software is compatible.
   - Make sure that your home computer and the printer at work are compatible. You can suffer enormous frustration if you discover incompatibility at a late stage.
   - Don't rely on printing your work at the very last minute. Murphy's Law operates at maximum level when you combine the last-minute principle with anything to do with a computer.

Checklist: planning and writing up a major report

Before you write any text

- Stop the practical work at a point that will allow you plenty of time to write up. It will take far longer than you expect.
- Plan a broad outline. Use Chapter 7 **How to structure a report** and scale up the processes.
- Draw up a preliminary Contents page.

Writing the first draft

- Start with the easiest section.
- Treat each section as a miniature story.
- Write from the perspective of the needs of your 'second marker'.
- Note the references you will cite, and where in the text.
- Plan the illustrations.

If you are using a computer

- Save and backup frequently.
- Make sure that your different computers and printers (e.g. home and work) are compatible.

Procrastination and Writer's block

- Accept that they are normal.
- Have strategies for dealing with them.

SECTION 3

The tools of technical writing

Introduction: the tools of technical writing

This section brings together all the basic tools needed for effective writing. It is meant to be used in conjunction with each of the sections **Writing an essay** and **Writing a report**. The topics covered are:

Doing the background reading and making notes

10 CHAPTER

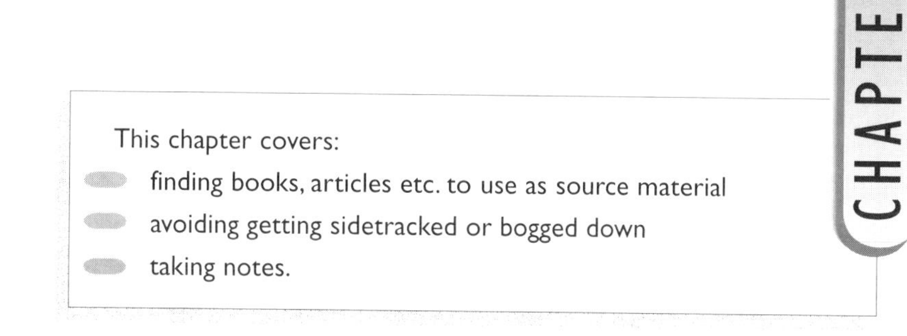

This chapter is particularly relevant to the writing of essays, and the type of report that does not involve original experimental or developmental work done by you or your group.

Writing a report or an essay in science or engineering involves responding to what others have written about the subject. Lecturers – who often call this detective work 'Reviewing the literature' – will want to know that you can find relevant material, review it and incorporate the facts and ideas into a reasoned study of your own.

The two areas where you can expect to have to do background reading as an undergraduate in the sciences or engineering are:

1 Reading for an original research or developmental project

A project that involves an element of your own original work has to show that you can assess your results, form your own conclusions and defend them, all within the framework of the other major work in your area. This involves searching for and identifying relevant source material, and then being able to assess your work in relation to that of other people. A high degree of organisation is needed.

Your supervisor will probably have given you original articles on your area of work; these will provide you with a number of leads into finding other literature. However, use this chapter if you need an introduction to the methods of finding source material in libraries, and taking notes. Then use Chapter 11 **Keeping records of work on a major project** for suggestions for organising a large amount of reference material.

2 Reading for an essay, or a report that is an analysis or description of an area of knowledge

This type of assignment involves reading around the subject and then analysing, interpreting and even challenging material from other sources. This chapter suggests ways of finding relevant books and articles, and how to extract information and take notes from them.

Step 1: Choosing and analysing the question

If you are writing an essay, the first, essential step is to analyse your chosen topic so that you know what sort of material you need to look for. Chapter 1 **Choosing an essay question** and Chapter 2 A**nalysing an essay question** will give help in these two areas.

Step 2: Finding relevant source material

It is often confusing knowing where to start looking for the relevant books, periodicals, journals and newspapers. There are a number of strategies that you can use.

Start with the recommended text book for your course

Read any chapters that may be related to the topic. If it has a Bibliography, look in it to see if there are other sources that are relevant.

Use the reading lists

A reading list is often given out along with the essay questions or the report topics. This contains the sources that your lecturer regards as the important background material. Be aware though, that he or she will expect you to find more sources than just these and the recommended text book for the course. The reading list provides an obvious starting point.

Hints from lectures

Reading material is often discussed in tutorials or lectures. Don't rely on remembering these vital snippets; make notes of them.

The library: familiarisation

Most people feel intimidated by an unfamiliar library. If your library offers a familiarisation tour at the beginning of the year, use it to grasp the basics of library searching. If you have missed out on it, or are still unsure, ask a librarian to show you the fundamentals. However, don't rely purely on the librarians to find your material. Their patience might begin to wear thin if you don't appear to be making genuine attempts to find things for yourself.

Strategies for library searching

1 **Use the desk copy material, then move on.** General texts will often have been put on desk copy. These will give you an

overview of the subject. Check the publication dates, and start with the most recent one. Journal articles tend to be more detailed, and provide the most up-to-date material on the topic.

2 **Use leads from books and articles.** While you are consulting the most important works on the subject, follow up their references and bibliographies. This will give you leads into other areas and perspectives. However, it's all too easy to become overwhelmed or sidetracked. The section below, **Avoiding getting sidetracked and bogged down,** gives ideas about how to stay afloat.

3 **On-line catalogues and CD-ROMs. An on-line catalogue** is a central computer data base, accessed from keyboard/monitor systems in the library. It allows you to search for books in your university library system by author, title, keyword or classification (call) number. Ask a librarian to teach you how to use them. It is easy to only tinker with them and not use them to their full potential. Used effectively, they are very powerful tools for finding relevant material.

   CD-ROMs are data bases on compact discs, kept in the individual libraries. They can be searched in the same way as on-line catalogues. Their information refers predominantly to specialist journal material.

4 **Use the Dewey classification (or whichever classification system your library uses).** Note the Dewey Classification Number (or other similar systems, such as the Universal Decimal Classification system (UDC), or the Library of Congress classification) of the books you have been recommended. Then check the on-line catalogue or the library shelves for books having the same number.

Explanation of the Dewey system

This is an international system for classifying and arranging the books in a library. Other classification systems work on similar principles. Books are divided according to subject matter into groups, each given a specific code-number. Principal subdivisions within each group are coded by adding further numbers, and with the use of decimal numbers further subdivisions can be generated without limit.

For example:

574	Life sciences
574.5	Ecology
574.52	Specific relationships and kinds of environments
574.526	Specific kinds of environments
574.526 3	Aquatic environments
574.526 32	Freshwater environments
574.526 322	Lakes, ponds, freshwater lagoons
574.526 323	Rivers and streams
574.526 325	Wetlands environments

Say, for example, you are writing about some aspect of freshwater ecology, and your deskcopy texts have the Dewey classification of 574.526 323 (Rivers and streams). In searching for more material, you could look at books on the shelves having the Dewey classification on either side of that subdivision: 574.526 322 (Lakes, ponds, freshwater lagoons) and 574.526 325 (Wetlands environments). To do a broader search, you could look at the books with a lower order of subdivisions, such as 574.526 3 (Aquatic environments).

5 **Don't forget to use encyclopaedias, yearbooks, handbooks, collections of statistics and newspapers.**

6 **Beating the rush.**There will be pressure on the library's resources for the material on the reading list. Competition can be fierce for desk copies, and even for shelf copies as the deadline draws near. It pays to get in early.

Material from companies and organisations

You can obtain a lot of useful source material from reports and publicity documents produced by companies, organisations and government departments. People employed by them are often pleased to help by giving information about their work and products. Don't be afraid to phone the receptionist and ask for published material, and perhaps to speak to someone who knows something about the subject.

Avoiding getting sidetracked and bogged down

It is all too easy to spend a long time reading around a subject and end up with little useful material, and also to feel overwhelmed by the sheer volume of published work in the area. It is very important, then, to avoid getting sidetracked. Use the following strategies:

1 Keep your topic firmly in mind

You can often find yourself reading something because it is intriguing you, not because it is relevant to your topic. Self-discipline is essential while doing a literature search. You need to think constantly of your topic. One strategy might be to have it written out in front of you while you read.

2 Skim-read your material

Reading reference material doesn't involve starting at page 1 and methodically reading the whole of it. A new reference needs to be skim-read to decide what degree of attention it deserves. There are several ways that you can survey material:

- **Book**: look at the contents page, chapter summaries, introductions and index.

- **Journal article**: read the Abstract or Summary.
- **If you find something that looks as if it may be relevant**, before you take notes skim the whole article or section by:
  a reading the first and last paragraphs
  b then skimming down the section headings
  c then reading the first sentence of each paragraph. This sentence should be a summary of the main theme of the paragraph.

If it doesn't look useful, be firm. Don't be dragged into reading parts of it because they look interesting.

Taking notes

Once you've decided that a reference is relevant, there are two ways in which you can record what is important:

- by writing notes
- by photocopying the relevant pages. But make sure that you don't haphazardly photocopy things just to feel you've achieved something.

Organising these can be a problem if you are writing a major project. Chapter 11 **Keeping records of work on a major project d**eals with this.

The following steps are useful strategies for note-taking.

Strategies for note-taking

The following strategies may strike you as being needlessly meticulous. However, students who are experienced researchers have found that a few minutes spent recording these sorts of details at the beginning of a search can save hours of frustration at the editing and revising stage.

1 **Write on one side of the page only**. This allows you to cut up and sort your notes later into a logical order.

2 **Record the bibliographic details.** The first thing to do is to **record the details that you will need later to construct your References section or Bibliography:** the author's name, the title of the article or book, and all the other necessary details. Chapter 13 **References** describes the information needed. It is essential to record all the details at the time of reading; you don't want to waste time coming back to do it again later.

   It is also a good idea to note the **name of the library** and the **call number**, so that you can find the reference easily again.

3 **If you are writing notes:**

- **Note-taking is time-consuming**. Keep asking yourself: is this relevant to my assignment? While you write, analyse the material, judge its usefulness, sift and classify it. Don't copy blindly.
- **Make a note of the source**. Every time you make a new note, write down the page number of the source. There are two reasons for this:
  - a In your assignment you will need to acknowledge each of your sources. Not to do so is plagiarism, and is serious. Chapter 13 **References** explains the importance of this and shows how to deal with citing sources.
  - b You may need to go back and find it again, to check something or get new material. The time-wasting that results from not having noted the page number is utterly frustrating.
- **If you want to quote a passage word-for-word,** make sure you copy it down meticulously. The words and punctuation of a direct quote in an assignment must be exact. The ways of dealing with quotes is given in Chapter 13.
- **Record on each of your notes the subject area(s)** that each will contribute to. This will help you sort them later.
- **Ideas will come to you while you make notes.** You will find that the process of assimilating other people's ideas while you make notes will give you ideas and perspectives of your own. Jot these down while you write, but make sure that you will realise afterwards that they are yours, not the author's. Tag them so that they are unique: circle, colour or highlight them.

4 **If you are photocopying the relevant pages:**

- **Mark the relevant passages** immediately after making the photocopy. It is easy to see the relevance of the original and then forget to mark the photocopy; this wastes time later.
- Don't forget to **record the bibliographic details** on the photocopied pages. You can waste a lot of time later if you forget to do this.
- Make a note on the photocopy of the **subject area(s)** that each covers. This will be useful later when you sort your notes.
- Also note **the date** you photocopied it. This gives you an understanding of how each article led to the next.

5 **Make up lists for each subject area.** Then cross-refer records of each reference to one or more of the lists.

When do I stop reading?

It is all too easy to procrastinate – to put off the writing process because it is always easier to keep on reading. At some point you have to decide to stop. Read through all your notes, keep your assignment question firmly in mind – or better still, have it written out in front of you – and consider whether you have enough information. If you have, then stop. Start planning your report (Chapter 9) or essay (Chapter 3).

Checklist: doing the background reading and making notes

- Start with the recommended text.
- Use the reading lists.
- Take hints from lectures.

In the library:

- Use the desk copies then move on.
- Take leads from Bibliographies and References sections of books and articles.
- Use the on-line catalogues and CD-ROMs.
- Use the decimal classification system.

Avoiding getting side-tracked and bogged down

- Keep your topic firmly in mind.
- Use strategies for skim-reading.

Taking notes

- First record the bibliographic details of the source.
- Select your points – don't copy blindly.
- For a word-for-word quotation, copy it meticulously.
- Note the subject areas to which each source is relevant.

Photocopying

- Record the bibliographic details of the source.
- Mark the relevant passages on the photocopy.
- Note the subject areas to which each source is relevant.

- **Make up lists for each subject area.** Then cross-refer records of each reference to one or more of the lists.

Chapter 11 Keeping records of work on a major project

This chapter covers the keeping of records on:

- work-in-progress
- computer runs
- software modifications
- photographs
- references
- material for a team report.

Efficient systems of record-keeping make your final write-up less traumatic. In a project that requires a major written report, such as a final year research project, you can end up by having a large amount of recorded material – written records of your experimental or developmental work, and the associated computer outputs, photographs, digital images, and records of reference material. At the beginning of the project, when not much has accumulated, they are all easily managed. However, if you don't put efficient systems in place, by the end of your project you can be floundering in a sea of records. It pays to start off with a system that you know is efficient and can cope with an ever-increasing amount of material.

This chapter gives suggestions about organising your material.

Work-in-progress: written records

There is one principle that is more important than any other in the area of keeping work records:

> Keep very detailed records of everything you do. Never assume that you are so thoroughly familiar with a procedure that you'll never forget it.

Common comments from students after the event are:

I wish I'd kept better records of that procedure I knew so well at the beginning of the year. I thought I'd never forget it. Then I had to repeat a few things. It took a lot more time than it should have.

Writing up was tricky. My records weren't good enough.

The two good reasons for keeping really detailed records are:

1 At the end of your project, when you are consolidating all your work, you realise that you need to fill a gap in your information and you have to repeat a procedure that has now faded from your memory.
2 When it comes to writing up, you've forgotten the procedure and won't be able to describe it in enough detail.

This aspect of keeping records can't be emphasised too much. It is so easy to fall into the trap of having inadequate records and so easy to avoid it. Reducing part of your detailed information when you write up is far preferable to having too little.

What is the best way of keeping written records?

Here are suggestions for keeping written records of work-in-progress:

1 **Have a hard-backed notebook:**

- Use it every time you work on your project. Loose sheets of paper are easily lost.
- Don't rely on making records of your work later. Two things could happen:
  a You'll have forgotten the small details, even by later on the same day.
  b It will never get done.

2 **The two-tier system:**

- **Use one book as a diary** to record every possible detail as it happens.
- **Use another book – or a word-processor – to write up as you go along.** As soon as you finish one phase of your work, write up the procedure and results in detail, and the discussion points that you think of at the time. **This is more than keeping records: this is a preliminary write up.**
  Two good reasons for it:
  a There is less to write when you get to the end of your research
  b It focusses your thoughts. Some student comments are:
  I used to try to keep it all in my head. But I couldn't think through issues well. Now I think through issues by writing. When the fog hits, I write.
  Writing is a great tool for helping you keep on track.

3 **Using a computer to keep records:**
This is only worth doing if you consistently have a computer available.

- **Spreadsheets** If you have repetitive experiments to do, it is worthwhile taking the time to customise a spreadsheet so that you can use it to record the date, the purpose of the experiment, components, quantities, temperatures, times etc. By using tickboxes, records can be entered very quickly.

- **Wordprocessors** These are powerful tools for anything following the first, written stage of on-the-spot record-keeping. They can make you feel that you are making progress; moreover, with some file management and editing much of your write-up is already done.

Records of computer runs and their associated outputs

If you are doing a number of computer runs where the input data are changed, it is all too easy to forget to record the input parameters corresponding to each output. If your software does not allow on-line recording of the data, it is essential to:

1 **Identify every run in a logbook** in a format such as:

Run number	Data	Comments
1 2 etc.		

2 **Record the run number and date on the output** as soon as it has been run.

3 **Have some sort of filing system for the printouts**. The pile-in-a-drawer system falls apart when a critical mass of printouts is reached.

Records of software modifications

More than any other scientific or technological area, software development can take tinkering to an art form. Unless the approach is very professional, much of the work often proceeds through a series of inspired hunches, and as a result it's very easy to forget the thought processes that led to a piece of code.

Comment throughout your code:

- **comprehensively**
- **in detail**
- **frequently**, not just when you think you're making a major change.

This will not only show professionalism on your part; detailed commenting will also:

- Help you to modify code that doesn't work, because you'll be able to understand the thought processes that led you to it.
- Make writing your report much easier, because you can understand the intellectual processes you went through.

- Help to make your programming procedures more rigorous.
- Reduce the frustration felt by other students who may have to develop your program for computers further the following year.

Saving and backing up

Students working with computers all have horror stories resulting from the lack of frequent saving and adequate backing up. False keystrokes, the system going down, an unexpected characteristic of an unfamiliar mainframe, a power outage, theft of a car containing the final write-up and the discs – all of these have wiped out lengthy pieces of work. Without backups, this is catastrophic.

With any work on a computer:

- **Save and backup frequently and thoroughly.** You can't be too fussy. Keep saving frequently while you are working. Systematically back up everything at the end of each day.
- **Keep backups at different places**, for instance, a set at home and another at work.

Records of photographs

Photographs need determined record-keeping because they're not immediate; you usually have to wait some time before the negatives are produced and even longer for the photographs.

Here are suggestions for keeping track of negatives:

1 **Record the frame numbers directly into your workbook**, together with any comments you want to make.

2 **As soon as you've developed the film or received it back from the photographic firm:**
- **Label it.**
- **Check that each frame number on the film corresponds with what you've written in your workbook.** A mistake with the frame numbers can usually be worked out by looking at the sequence. But remember that images such as transmission electron micrographs – memorable while you were working on the instrument – can look confusingly similar in negative form.

3 **File your negatives or colour slides** in a system that will protect them and will allow you to identify each frame easily.
- **Negatives.** Their protective pockets can be labelled and stapled into a folder.
- **Slides.** Each slide should be numbered and information about each one recorded elsewhere, with careful cross-referencing to the numbers.

4 **Make contact prints of all black-and-white negatives** (they don't have to be perfectly exposed) and stick them into a folder or book.

5 **Don't throw any away**. They may look unimportant or imperfect at the time, but you may find they'll become significant after further work.

Records of references

1 **Specialist software for keeping records of references.** If you know you're going to have to deal with more than about 50 references it is worthwhile using specialist referencing software or customising a data base. These will allow you to search by keyword, and to customise the formatting of your output.

Even if at first it looks as though you won't have many references, it's as well to remember that you may end up having to deal with more than you first thought, and that your system is going to have to cope. Some suggestions are:

2 **Photocopying references.** If you photocopy reference material, remember when you first read it to:

- a Either give each one a number in the order that you acquire them *or* file them in alphabetical order of the first author's surname.
- b Make a note on the front about the different areas covered in each paper. For instance you may have a paper that deals principally with uptake of heavy metals, but also mentions dioxins and aspects of the measurement of water quality. Note all three.
- c Keep separate lists of these different areas of interest, and note the reference number or author(s) and date of the various papers. One paper can therefore appear on more than one of the subject area lists. The lists can be on computer or paper.

3 **Keeping written notes of references:**

- a The most efficient option is to use a separate sheet of paper for each reference.
- b Make sure you write down all the details you're going to need when you cite the reference in your report (see Chapter 13). When you write up, a lot of time can be wasted having to find out small details you forgot to note, such as the final page number of a paper or the editor of a book.
- c Write a brief summary of all the points you think are going to be useful to you.
- d Then go through stages a-c, **Photocopying references**, above.

Records of a team report

To write a successful team report needs careful documentation of each individual's effort.

Doing the hands-on work is only part of the story. Each person should also keep a record of the amount of time taken in:

- thinking
- asking questions
- interacting
- meeting
- reading
- typing
- correcting
- collecting feedback.

You may want to choose a Group Administrator for the assignment in addition to the usual Coordinator. The function of the Administrator would be to:

- keep concise minutes of each team meeting, with a stress on conclusions and major actions arising from the meeting
- gather at regular intervals details of the number of hours worked by each group member
- facilitate the writing of the report.

You may be asked to submit a private assessment of the contribution to the project of every team member, including yourself.

Checklist: keeping records of work on a major project

- **Keep *very* detailed records of everything**. This is essential to avoid wasting time on:
  - ✓ repeating a procedure when you've forgotten the details
  - ✓ doing the final write-up.
- **Have a hard-backed notebook**. Don't use single sheets of paper – they get lost.
- Use a two-tier system:
  - ✓ a notebook for day-to-day work
  - ✓ another notebook or a computer to write up as you go along. Use this to keep on track.

Using a computer

- Use spreadsheets for details of repetitive experiments.
- Keep systematic records of:
  - ✓ computer runs and their outputs
  - ✓ software modifications: comment frequently in the code.

Photographs

- Record frame numbers in the notebook.

- Cross-check them with the developed film.
- File negatives and slides: cross-refer numbers and information.
- Make contact prints of black-and-white negatives.
- Don't throw any away.

References

- If a lot – use specialist software. Otherwise:
  - ✓ keep records of each reference on a separate sheet
  - ✓ make sure all the bibliographic details are recorded
  - ✓ make up lists of the subject areas: then cross-refer records of each reference to one or more of the lists.

A team report

- Each person should keep detailed records of their area of responsibility and the interaction.

Illustrations

CHAPTER 12

This chapter covers:

- whether to use a graph, table or text
- deciding which type of graph to use
- drawing up informative graphs and tables
- labelling them correctly
- numbering illustrations and referring to them in the text
- drawings, diagrams and photographs
- using digital images in the final printout.

Technical writing relies heavily on presenting information visually. There are two aspects to it:

1 **The illustrations themselves.**

2 **The way that they are labelled and referred to in the text.** This is an area where it is easy to lose marks, even when the graphs and tables themselves are well thought out.

This chapter gives a brief overview of the conventions for the commonest types of illustrations for university assignments:

- line graphs
- bar (column) charts
- pie charts
- tables.

A graph, a table or text?

Think about:

- How can the subject matter be most clearly presented?
- How are the most important points in the data best clarified?
- How much detail and accuracy are required?

Example:

Here are three ways – text, table and graph – of presenting the same information: the rise and fall of temperature over time, measured at irregular intervals:

Text

t = time, T = temperature
$t_1 = 0.00$, $T_1 = 6$°C: $t_2 = 9.55$, $T_2 = 14$°C: $t_3 = 12.10$, $T_3 = 16$°C: $t_4 = 17.50$, $T_4 = 23.5$°C: $t_5 = 20.15$, $T_5 = 25$°C: $t_6 = 25.00$, $T_6 = 20.5$°C: $t_7 = 34.50$, $T_7 = 19$°C: $t_8 = 40.00$, $T_8 = 7$°C:

The pattern is difficult to decipher from these data.

Table

Time	Temperature (°C)
0.00	6
9.55	14
12.10	16
17.50	23.5
20.15	25
25.00	20.5
34.50	19
40.00	7

It is obvious that the temperature peaked, but the pattern of the rise and fall is not clear. The exact times can be seen, but are they necessary?

Graph

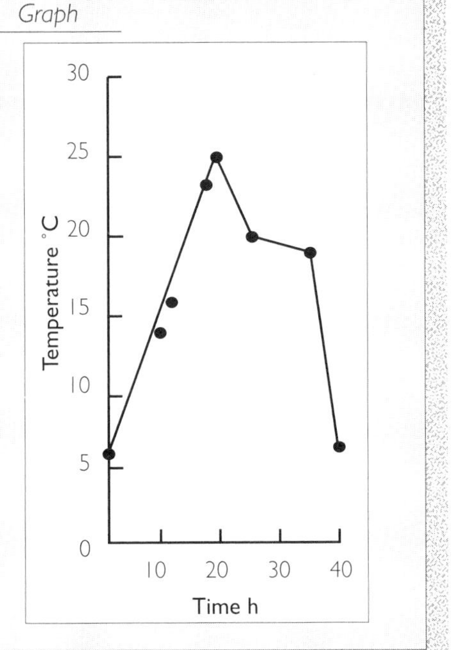

Here it is clear that the rising phase was linear, but not the falling phase.

Use a table

- where **many numerical data** have to be presented in a **logical** way
- if **exact numbers** are important.

Use a graph

- for any data that show a **trend**
- for **more immediate impact** than a table (tables have to be interpreted).

Graphs also help to compare:
- experimental results and theoretical predictions
- results from different sources
- the size of errors with the size of expected effects.

Which type of graph?

- Use a **line graph** to depict trends or relationships.
  - **a** In a ***trend***, the same data change over time (e.g. the rainfall of one area at different points of time).
  - **b** In a ***relationship***, there is an interaction between two variables (e.g. load versus extension, or respiration versus temperature).
- Use **semilog** or **log-log** plots instead of a linear line graph when your data need compression or when they obey certain types of relationships, see page 94.
- Use a **bar graph** to compare discrete items (e.g. the rainfall of three different areas at one point of time) see page 95.
- Use a **pie chart** to represent separate parts of a whole (e.g. types of land use in a country) see page 96.

Guidelines for plotting the main types of graphs

1 Linear line graphs

A The axes

- Use the x (horizontal) axis for the independent variable (the one that changes automatically, particularly time).
- Use the y (vertical) axis for the dependent variable (the one you measured).
- Choose the width of your scale, taking into account the largest and the smallest numbers for each axis.

Remember that your viewer's emotional perception of your graph could be different depending on the width of the scale. By choosing an inappropriate scale you can:

- create an uninformative graph, because it's too squashed up or
- look as though you're obscuring trends or emphasising them too much.

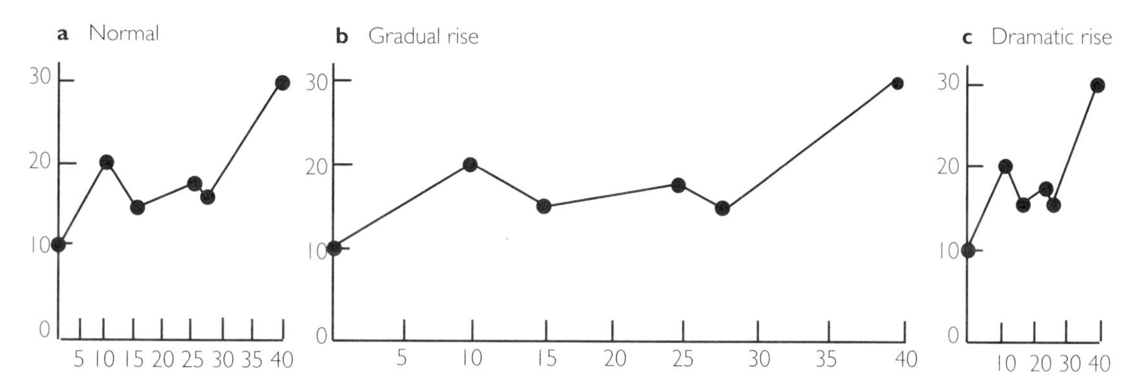

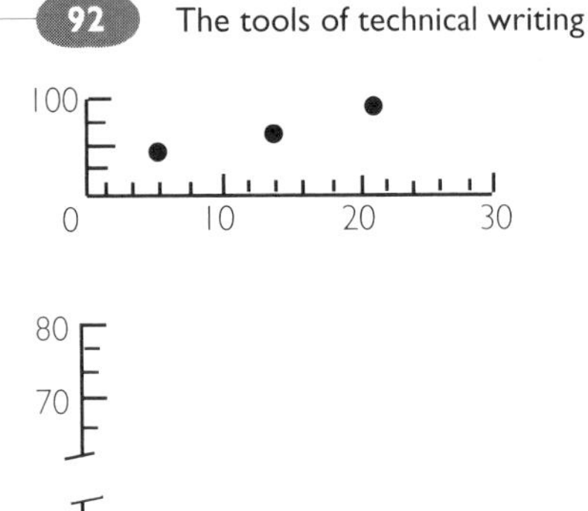

- Choose axis marks that are 'main' or 'round' numbers (0, 2, 4, 6... or 0, 10, 20, 30...), no matter if your data points are intermediate to these.

  For example, the values of the data points would be difficult to estimate for an axis labelled 0, 7, 14, 21…

- If the origin of an axis does not need to be zero, show a break in it.

B Drawing the graph

- If you are using graph paper: **leave enough room to label the axes.** Most commercial graph papers have margins that are much too small for adequate labelling of the axes.

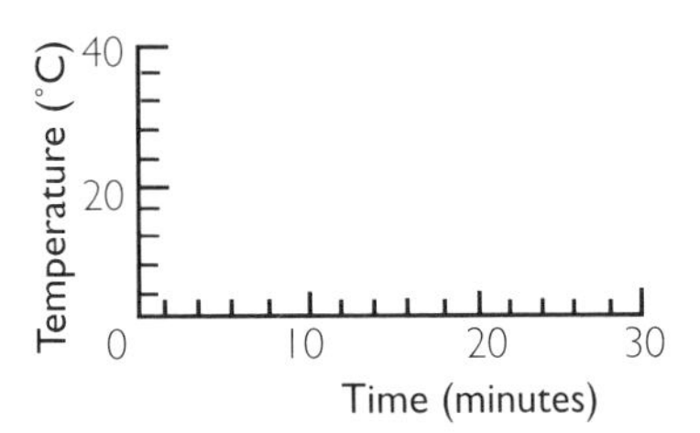

- **Number the tick marks along the axes.** If the numbering looks too crowded, leave out some of the intermediate values. Do it consistently, e.g., label the tick marks 0, 10, 20… not 0, 5, 15, 20.

- **Label both axes clearly. Don't forget to include the units.** Many people forget to include the units and lose marks for it.

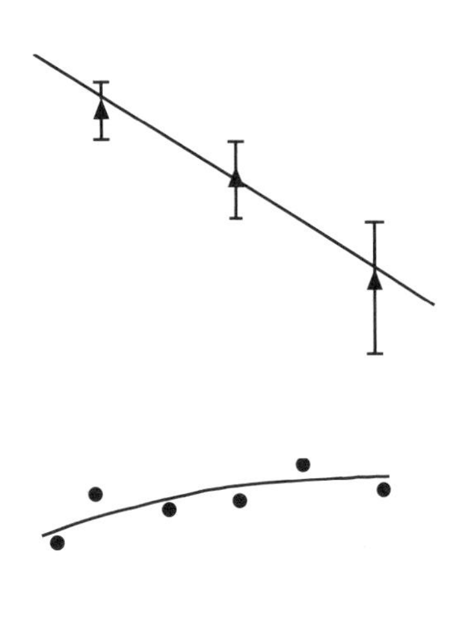

- **Mark the data points.** Make them big enough to be seen clearly, at least 1 mm diameter, not just a barely visible pinpoint.

- **Error bars:** These are used to indicate the variability in individual estimations without including all the data. They are usually vertical bars shown on the independent variable, extending from -1 Standard Deviation (SD) or Standard Error of the Mean (SEM) to +1 SD or SEM, centred on the datum.

- **Draw a curve**

  a **When appropriate, join the points by a smooth curve.** If all the points do not lie on a smooth curve but it is clear that there is a continuous relation between the variables being plotted, draw a 'best-fit' smooth curve.

  Join the data points by straight lines if you have discrete points such as monthly rainfall data.

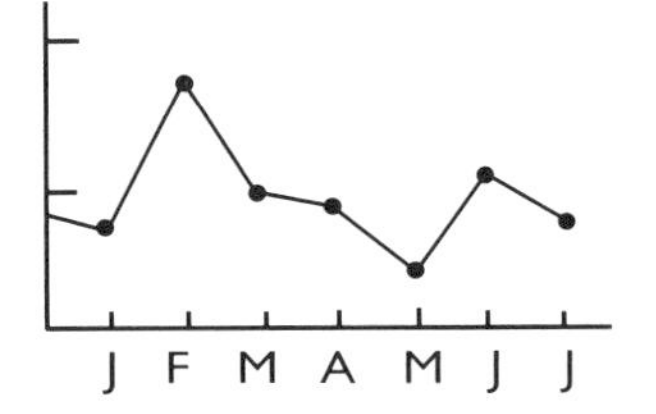

  b **If you are drawing your curve freehand, make sure that it is a single, narrow, smooth line.** Avoid making it wide and scratchy, or drawing wiggles that deviate from the main path.

  c **If a single measurement is in error** for a known reason, you can discard it. Do not discard it if you know of no clear reason for the departure from the expected pattern.

- **To differentiate between different lines,** use different symbols (squares, triangles, circles, open and closed versions) and different types of lines (solid, dashed, alternating dashes and dots).

- **Don't use too many lines.** Reduce the number of lines and put the data on another graph if:
  a you have two or more lines on one graph, and if they cross over too much, or
  b if the graph looks too crowded.
  Don't aim for the least number of graphs; the important principle is that each of your graphs should be easy to interpret.

- **Orientation of the graph:** if possible, graphs should be able to be read without turning the page on its side. If they have to be read from the side, the convention is that they should be read from the right hand side of the page.

C The title and legend

- Each graph **must** have a figure number and a title (e.g. Figure 1 Growth of *E. coli* cell culture over 24 hours) and a legend if needed.

- **The title** must identify the main point of the graph: e.g. Load-extension curve for 1 % carbon steel *or* Growth of *E. coli* cell culture in minimal media.

- The figure number and title should be written below it. This is a scientific convention: headings are placed *below* figures and *above* tables.

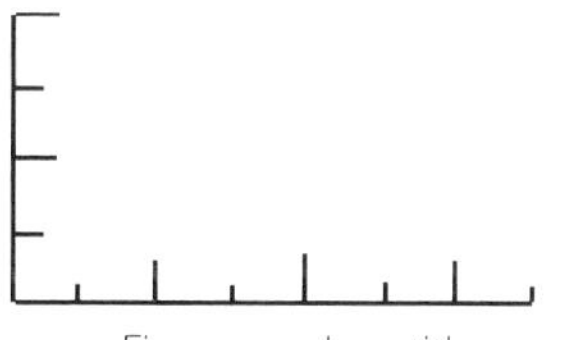

- **The legend:** a small amount of explanation of your graph can usually be incorporated into the title. For example: Growth of *E. coli* cell culture under optimal (o—-o) and stressed (x—x) conditions. A longer description must be incorporated into a legend; this follows the title and explains the figure or identifies its parts.

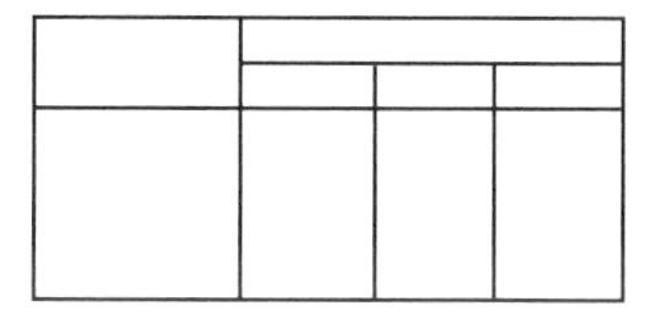

- Each figure should be self-explanatory. Your reader should understand it without having to look at the text.

> *Each figure must be referred to by its figure number at the appropriate point in the text.*
> *See page 99 for how to do this.*

2 Semilog and log-log graphs

Two types of graph paper other than linear are commonly used for plotting data:

- **semilog** – where one axis is linear and one logarithmic
- **log-log** – where both axes are logarithmic.

The reasons for using logarithmic scales are:

- The data may cover a large range and need to be compressed.
- A log plot may arrange nonlinear data in a straight line.

Calibration of log graph paper

The figure below shows (a) a single logarithmic scale and (b) a three-cycle logarithmic scale. When using multi-cycle graph paper, use the one with the fewest cycles needed to represent your data.

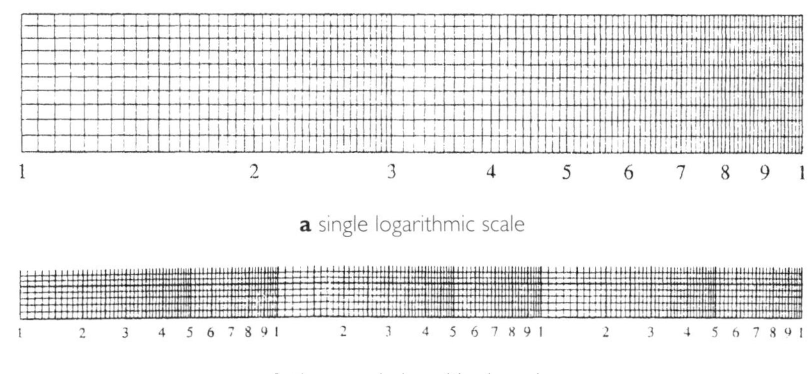

a single logarithmic scale

b three-cycle logarithmic scale

Why does using log graph paper produce a straight line from nonlinear data?

A Using semilog paper

Consider the following equation:

$$y = be^{mx} \tag{1}$$

Taking the log of both sides, we have:

$$\log y = mx(\log e) + \log b \tag{2}$$

In equation (2), log b and $m(\log e)$ are constants. Therefore the equation is in the form of a straight line, with log y on one side and x on the other.

Data that are described by equation (1) would plot as a straight line on semilog graph paper.

B Using log-log paper

Suppose your data are represented by a curve obeying the relationship:

$$y = bx^{m} \qquad (3)$$

Taking the log of both sides, we have:

$$\log y = m(\log x) + \log b \qquad (4)$$

In equation (4), log b and m are constants, but the two variables are represented as log values. Therefore the equation is in the form of a straight line, with log y on one side and log x on the other.

Data that are described by equation (3) – that is, would describe a curve when plotted on a linear scale – would plot as a straight line on log-log graph paper.

3 Bar (column) graphs

Description

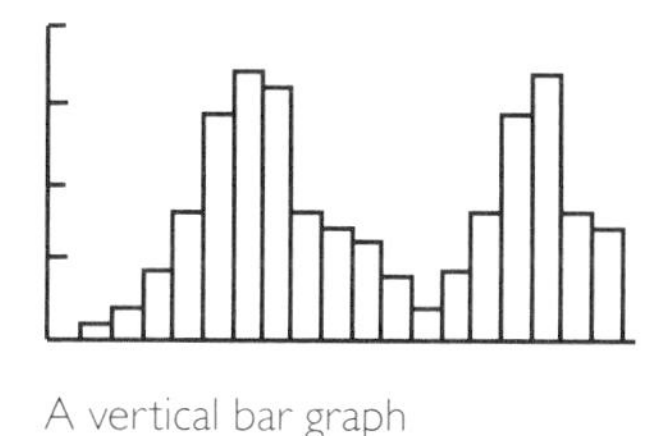

A vertical bar graph

- A **bar or column graph** uses rectangles to indicate the relative size of several variables, thus comparing the items by means of the height or length of the bars.
- Bar graphs can be either horizontal or vertical, depending on whether the bars go up the page or across it.

When to use

Vertical bar graphs (often called column graphs) are usually used for showing discrete values over time. Either type can be used in most other instances.

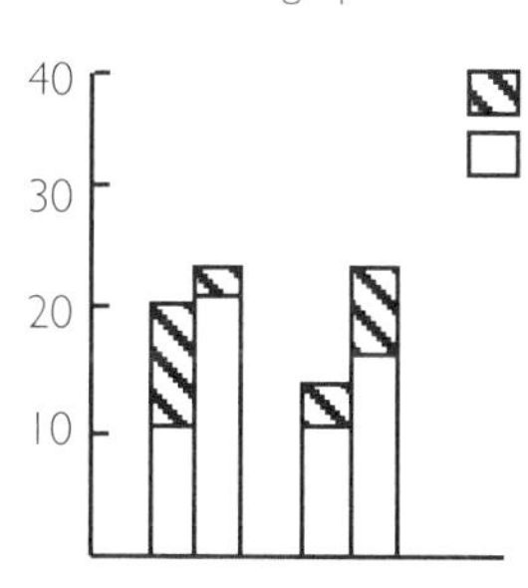

Subdivided bars

Guidelines for constructing vertical bar graphs

- Place the names of the items you are comparing – the independent variable – under the bars.
- Place the units of comparison – the dependent variable (usually numbers) – at the left.
- Use a legend – a small sample of the markings and brief text – to explain the meanings of the bars' markings.
- The bars on a bar graph can be subdivided to show relative proportions.
- Rearrange the items accordingly for horizontal bar graphs.

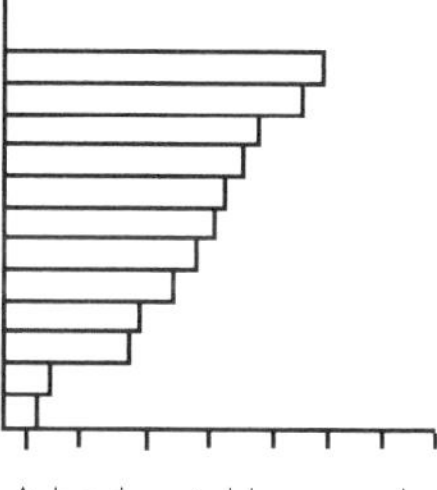

A horizontal bar graph

4 Pie charts

Description

A pie chart is made up of a circle divided into segments.
Its commonest use is to illustrate relative percentages of a whole.

When to use

When dealing with percentages. A pie chart lets you compare the size of the individual parts with each other and with the whole.

When not to use

If there are too many segments and if they are too small, the labelling becomes impossible. Consider presenting the data in another way, perhaps as a table.

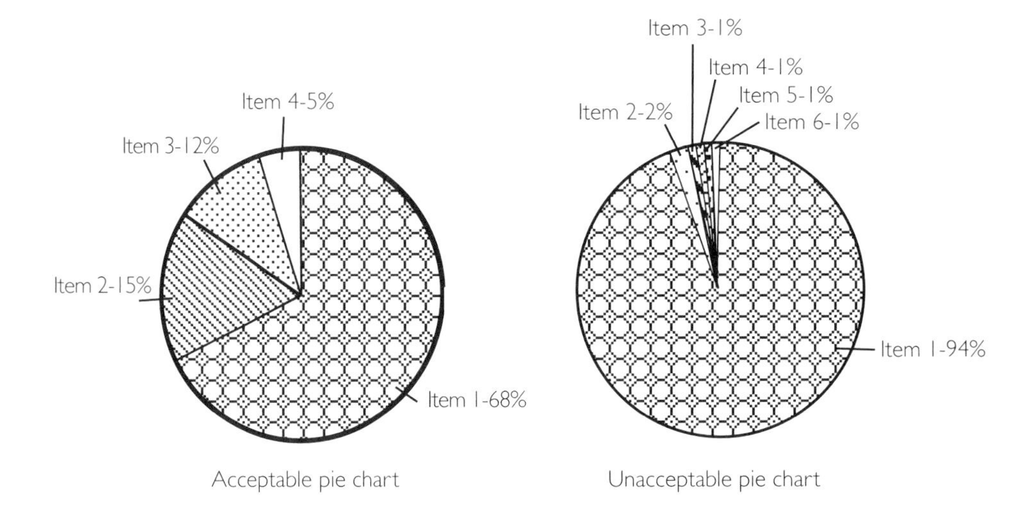

Acceptable pie chart

Unacceptable pie chart

Tables

Description

Tables present information – usually numbers, but sometimes words – in columns and rows.

When to use

To show classifications and relationships of numerical or verbal data.

Guidelines for preparing tables

- Put the items you want to compare (the independent variables) down the left hand side of the table.

- Put the categories you are comparing (the dependent variables) across the top in the column headings.
- All columns must have headings. These should include the units and any scaling factors used.
- Use a spanner head to name the column heads below it. Spanners reduce repetition in the column heads. (See Table 12.1)

- **Important:** columns are easier to compare than rows, since it is easier for us to run an eye down a column to compare data than to run them across. In Table 12.1, it is very easy to see by running down the columns that reindeer milk has by far the highest protein, fat and total solids content, and the lowest lactose content, of all the milks listed. This is not obvious from Table 12.2, where the data in the rows and columns have been transposed.

Table 12.1 Composition of milk from various mammals
(From: *Biotechnology*, Vol. 5: Food and Feed Production with Microorganisms, ed. Reed, G, Verlag Chemie, Weinheim, 1983)

Spanner heading
Column headings (dependent variables)
Row headings (independent variables)
Column
Row

	Proximal analysis %			
Mammal	**Protein**	**Fat**	**Lactose**	**Total solids**
Cow	3.40	3.72	4.90	12.74
Goat	3.20	3.90	4.50	12.40
Mare	2.49	1.59	5.90	10.38
Water buffalo	4.00	7.98	5.18	17.95
Sow	6.00	6.85	4.90	18.70
Camel	3.50	3.50	5.00	12.70
Reindeer	11.46	16.90	2.75	32.54
Sheep	6.00	7.00	4.50	19.00
Dog	7.10	8.30	3.70	20.40
Human	1.30	3.70	7.00	12.20

Table 12.2 Composition of milk from various mammals. The same data as in Table 12.1 are presented differently, in that the rows and columns are reversed.

Mammal	Cow	Goat	Mare	Water buffalo	Sow	Camel	Reindeer	Sheep	Dog	Human
Protein %	3.40	3.20	2.49	4.00	6.00	3.50	11.46	6.00	7.10	1.30
Fat %	3.72	3.90	1.59	7.98	6.85	3.50	16.90	7.00	8.30	3.70
Lactose %	4.90	4.50	5.90	5.18	4.90	5.00	2.75	4.50	3.70	7.00
Total solids %	12.74	12.40	10.38	17.95	18.70	12.70	32.54	19.00	20.40	12.20

- Plan your table so that there is adequate spacing of columns and to avoid splitting the table across two pages.

- Any table too wide to fit upright on a page should be presented so that it is read from the right-hand side of the page.
- Each table must have a **table number** (see page 100) and a **title**. It may also need explanatory notes placed under the table. Number these by superscripts: [a b c].
- **Put the number and title above the table.** This is a scientific convention: headings are placed *below* figures and *above* tables.
- Large tables and those not essential to understanding the main story of your report, but which are needed so that the detailed information is available, should be placed in an appendix (see page 62).

Drawings and diagrams

Guidelines for drawings and diagrams

- Each diagram should be directly relevant to the text. Don't just put one in because it looks nice.
- If you are drawing them by hand:
  - **a** Don't use a ball-point pen. It always looks scruffy.
  - **b** Use fine-tipped drawing pens or pencils.
  - **c** Make sure that your touch with a pencil is not so delicate that the line can barely be seen.
- Make each one self-sufficient, so that the reader does not need to look through the text for explanation:
  - **a** Label each one well.
  - **b** Give each one a good, explanatory title and a good legend (the writing that follows the title and explains the features of the diagram).
- If the diagram has been photocopied or cut out from another source it **must** be acknowledged. If you don't, you are implying that it is your own work; this is plagiarism (literary theft) and is serious. Chapter 13 explains this and gives guidelines for citing the source of a diagram.

Photographs

You may need to use photographs:

- To help your verbal description – showing characteristic fracture patterns, illustrating the species that you are working on, the topography of your study area.
- To help prove something you are asserting – a polluted waterway, foreshore litter.
- As a major element of your work, particularly work involving microscopy.

Guidelines for using photographs

- Do not use bad prints – in particular, prints that are of the wrong contrast and which hide detail.

- If you have to produce copies of your report, be aware that most photos reproduce atrociously when they are photocopied. You may have to stick photos in each copy of your report.
- For the more advanced form of report, many departments now require photographs to be scanned into the main text, and the originals to be inserted into the appendices. This is a measure to guard against electronic tampering. It is not so critical in undergraduate reports, but be aware that your department may have requirements in this area.

Numbering your graphs, photographs, drawings and tables, and referring to them in the text

Even if your figures and tables are well thought out and beautifully presented, it is possible – and very common – to go wrong in two areas:

- in the way that they are numbered
- in the way that you deal with them in the text.

Guidelines for numbering of figures and tables

1 **There should be two numbering series** – one for figures and one for tables.

2 **Figure numbers.**

All your figures – this includes graphs, line drawings, maps, photographs, and other types of illustrations – should be labelled as one series. Each successive one of the series is then labelled Figure 1 (*Title*), Figure 2 (*Title*) etc.

Example: If the first of your illustrations is a map, the second a graph and the third a line drawing, call the map Figure 1, the graph Figure 2, and the line drawing Figure 3.

Common mistake:

Distinguishing between the different types of illustrations and calling your graphs Graph 1, Graph 2..., your maps Map 1, Map 2... and your line drawings Figure 1, Figure 2....
The professional way to do it: Have them all as one series, each one of which is called a figure.

Variation on this: numbering according to the section number
When your report has numbered sections (see page 53), your illustrations can be numbered according to the section numbers:

First illustration	a map in Section 2	Figure 2.1
Second illustration	a graph in Section 2	Figure 2.2
Third illustration	a line drawing in Section 4	Figure 4.1
Fourth illustration	a graph in Section 5	Figure 5.1

Do not try to number your illustrations according to the sub-section number. You will end up with figure numbers such as Figure 3.1.2.1 (the first figure in Section 3.1.2), which is clumsy and confusing. Number them in accordance with the main section only.

3 **Table numbers.**

The numbering series for your tables is completely independent of the series for your figures.

Number each table Table 1 (followed by the title), Table 2 (followed by the title).... in sequence.

Variation on this

In the same way as for figures, tables can be numbered according to which section of your report they occur.

First table	the first one in Section 2	Table 2.1
Second table	the second one in Section 2	Table 2.2
Third table	the first one in Section 5	Table 5.1

4 **For illustrations in the appendices.**

Number them according to the appendix number, e.g. Figure 2, Appendix 1.

Referring to your figures and tables in the text.

Every figure (graph, line drawing, map etc.) and every table must be referred to somewhere in the text.

Common mistakes:

1 Not referring to some of the illustrations in the text.
2 Unnumbered and unlabelled illustrations, sometimes referred to in the text as '...as shown in the illustration below'.

At the relevant point in the text, refer to each figure or table in one of the following ways:

- At the beginning of a sentence: 'Figure 3 shows that the ...'.
- At the end of a sentence or phrase, immediately before the full stop or comma: '...that the rate of uptake increases exponentially (Figure 3).'
- '...that the rate of uptake increases exponentially (Figure 3), and that the ...'
- 'It is shown in Figure 3 that the ...'.

Do not write: 'In Figure 3 you can see that the ...'

Where to place illustrations

Illustrations should be placed as near as possible to the place in the text where they are first mentioned. However, large and complex illustrations such as detailed maps, geological cross-sections etc. may be better placed in an appendix. Every illustration you put in an appendix should be directly relevant to the text.

Using digital images in the final printout

There are two areas where you can expect to have to manipulate files of digital images:

- Computer-generated graphs and drawings.
- Digital images from instruments such as scanning electron microscopes and confocal microscopes.

Guidelines for dealing with digital images

1 **When you have only a few illustrations.**

- **Leave gaps in your text files.** Many people, if they are dealing with only a few graphs or illustrations, get into trouble when they try to embed these images in the text because they haven't the time to learn the correct file-handling techniques. If you have only a few illustrations, it is probably quicker to leave gaps in the printout of your text and cut-and-paste the illustrations with invisible sticky tape. A good photocopy of the page can then be used in the top copy of your report. Even allowing for the time needed to fiddle with scissors and tape, this is usually the most time-efficient way.

2 **If you have many illustrations and want to embed them in the text.**

Take advice on the best file-handling methods for your software package and hardware system. Some suggestions are:

- If possible, store the images on a hard disc, rather than having a number of floppies. However, they can be so large that unless your department has a server you may be constrained to using a lot of floppies. If so, remember to keep several backup copies, all in different locations.
- It is a great mistake, when you are writing up, to embed these images directly into your text files. The file size – particularly of microscope images – is so huge that your text files very quickly become unmanageably large. Editing such files then becomes clumsy. The most efficient method is to create a separate sub-directory for them, and then import each one into the text file. Then when you edit the text file, you can choose to suppress the display of the images.

Checklist: Illustrations

Use a graph:

- For any data that show a trend.
- For more immediate impact than a table.
- To compare:
  - ✓ experimental results and theoretical predictions
  - ✓ results from different sources
  - ✓ the size of errors with the size of expected effects.

- Use a **line graph** to depict trends or relationships (page 00).
- Use **semilog** or **log-log plots** instead of a linear line graph when your data need compression or when they obey certain types of relationships (page 00).
- Use a **bar graph** to compare discrete items (page 00).
- Use a **pie chart** to represent separate parts of a whole (page 00).

Use a table:

- Where **many numerical data** have to be presented in a logical way.
- If **exact numbers** are important.

Each illustration should:

- Have a Figure/Table number followed by the title.
- Have a title that is explanatory.
- Be understood without having to look at the text.
- Be referred to at the appropriate place in the text.
- Be placed as near in the text as possible to where it is first mentioned.

To embed digital images in the final printout:

- If there are only a few: leave gaps in the printout, stick them in and photocopy.
- If you want to import them into the file:
  - ✓ learn the appropriate file-handling techniques
  - ✓ do not embed them directly in the text. Create a separate sub-directory for them and import each one.

References

13

CHAPTER

This chapter covers:

- citing references in the text
- presenting a References section at the end of your report or essay.

It assumes you have no prior knowledge of how to reference.

This chapter deals with how references should be used in the body of a report or essay, and how to present a list of references or a bibliography at the end of your document. This is an area where you may feel inadequately prepared; it is one of the most convention-ridden areas of scientific and technological writing. Some lecturers take a relaxed attitude to the conventions; others expect them to be observed in the minutest detail. Many course notes give only bare outlines of how to do it.

No previous knowledge is assumed in this chapter. The aim is to give you all the skills required in this area.

Why you have to be exact when you document

It is essential that the sources of all your quotes and references are acknowledged in the accepted format. It is easy to get details wrong. Be aware that most lecturers are extremely meticulous about the way sources are referenced in assignments, and will check your work very thoroughly.

Documentation must be thorough because

- All scientific and technological work has to be put in the context of other work in the field. Your reader has to know that you are familiar with the literature in your area, and that you can assess your work in relation to it.
- Other people must be able to follow up the reference if they wish. This means it has to be cited accurately and in detail.
- Failure to acknowledge sources is plagiarism, or literary theft. Plagiarism is regarded very seriously. People who do not fully acknowledge their sources are copying the work of others and

implicitly claiming that the work is their own. They face ostracism and, sometimes, legal action and adverse publicity. Students risk failing their assignment, exclusion from their course and, sometimes, suspension from university.

When should references be used?

You need to use references:

1 **When you cite factual material taken from other sources** (see pages 105–113). This is the commonest form of documentation in a science or engineering undergraduate assignment. It is the form used almost exclusively when you write papers for journals, and is therefore the form that is monitored the most critically by lecturers.

It includes papers in professional journals, books, magazines, newspapers, an organisation's publicity material, engineering standards, government documents and legal statutes.

2 **When you need to quote word-for-word from another work** (see page 114 Using Direct Quotations).

Differences in documentation between arts-related subjects and the sciences

In arts-related subjects you may have written essays that required a footnoting system using *ibid*. and *op. cit.* to cross-refer to previously cited sources. This system is not used in scientific and technological literature.

References section or Bibliography?

- **A References section** is a list of all the sources that you have cited in the text of your document.
- **A Bibliography** is a list of all the sources you have consulted while writing your document, only some of which are cited in the text.

Most departments in science and engineering require a References section, because this demonstrates that you are familiar with the literature and can cite it appropriately in your own work. It is essential to find out what your department needs.

How to document using a References section

There are two interlinked aspects to this:

1 Citing each reference in the text.
2 Compiling the list of references for the section called References at the end of your document.

How to cite references in the text

There are two schemes commonly used in scientific and technological work for cross-referencing citations in the text and the full reference.

Most university departments prefer one or the other system. It is essential to find out what is the recognised method in the department for which you are writing. You should always use the system that your department prefers, and use it consistently throughout your report or essay.

Method 1: author's name and the date of publication (the Harvard System)

- **In the text**, citations consist of the surname of the authors(s) and the year in parentheses ( ). There are details of how to do this on pages 106–111.
- **References section:** the full details of each reference are then listed at the end of your assignment in a section called References, in alphabetical order of the first author's surname. Pages 107–111 show how to do this, and page 112 shows an example of this system.

Method 2: the sequential numbering system

- **Each citation in the text is given a unique number**, either in square brackets [ ] or superscripted, and numbered in the order in which they appear in the text.
- If you need to cite a reference more than once in the text, the number of its first appearance is used each time you cite it.
- **The References section** is made up of a sequentially-numbered list (i.e., not in alphabetical order as in the author-date system). Pages 101–111 show how to do this, and page 113 shows an example of this system.

Disadvantages of the sequential numbering system:

1 It is almost impossible to add another citation and renumber all successive ones without getting in a muddle. However, wordprocessors that automatically generate footnotes do help if you are required to use this system.
2 The numbers give no information about the work, and it is easy to forget to use the earlier number when you need to refer to it again later in your report.

3 A reader familiar with the literature cannot immediately recognise the piece of work you are citing.

For these reasons, the author-date system is often used in preference to the numbering system. It is essential to find out which one your department needs.

How to use the author-date (Harvard) system

Here are examples of how to cite the source in the text if you are using the author-date system:

Where the author's name has been quoted in the text:

> Mylona (1989) has analysed changes in sulphur dioxide and sulphate concentrations in air during the period 1979–1986.

Where the author's name has not been quoted in the text itself:

> The wind velocity and behaviour of a geographical region is a function of altitude, season and hour of measurement (Johnson, 1985).

Where the references need to be precisely placed:

> This runoff has also introduced heavy metals (Louma, 1974), pesticides (Schultz, 1971), pathogens (Cox, 1969), sediments (Gonzalez, 1971), and rubbish (Dayton, 1990).

Where the paper quoted is by two authors:

> Fatigue cavitation in grain boundaries has been reported to occur when the cyclic frequency is below a critical value (Woodford and Coffin, 1974).

Where the paper quoted is by more than two authors, cite the name of the first author and add 'et al.'. In some house styles, this is italicised.

> Lee et al. (1992) found no evidence of heather decline in nitrogen enrichment experiment on an upland heath in Wales.

Where several sources are quoted, separate them by semicolons, and cite them in order of publication date:

> Experimental addition of nitrogen to heaths in the Netherlands has caused a conversion of these communities to grassland (Heil, 1984; Berendse, 1990; Van der Eerden et al., 1991).

Where two or more papers written in different years by the same author are quoted:

> If the interfacial shear stress is assumed to be constant, the recovery length is related to the maximum shear stress in the fibre (Curtin, 1991, 1993).

Where more than one paper by the same author written in one year are quoted, distinguish between them by adding a lower case letter to each paper.

> Juvenile oysters attach their larval left valve to the substrate using fibrous organic material secreted from the foot (Cranfield, 1973 a, b, c). One of these studies (Cranfield, 1973b) also showed that......

An illustration copied from someone else's work:
The source of any illustration that you reproduce in your own report or essay must be cited. At the end of the caption below the illustration in your text, write (Reproduced from *author, date*).

> Figure 2.1 A typical graphite block heat exchanger (Reproduced from Hewitt, 1990).

Personal communications:
If someone has told you or written a note to you about an aspect of your work, it is quoted as *pers. comm*. Cite the initials and the surname. Note: Personal communications are not included in the References section.

> The sample was maintained at 25 °C and pH 5.0 (D.J. Wilson, pers. comm.).

The following example is unlikely in a report, but may occur in an essay with a historical component.
Where the publication date of the source is known approximately, use a small c before the date:

> All the branches of a tree at any degree of height, if put together, are equal to the cross-section of its trunk (Leonardo da Vinci, c. 1407).

Where you are citing a major source a number of times, it is useful to the reader if the individual page numbers are cited with each text reference:

> (Clarkson, 1995, p 51)

How to compile the References section

The References section is made up of a list of the papers, books, articles etc. that you have cited in the text of your work. It is placed at the end of your document, before any appendices (see Chapter 7).

Points to note:

- Each reference is listed only once.
- There are minor variations in the way the lists are cited for different house styles, as, for example, in the position of the date, the use of italics, quote marks etc. **It is important to find out exactly the form that your department requires**, and to stick to it rigidly.
- Be sure that every full-stop or comma is in the right place, and all other aspects of the formatting are correct. Formatting of references is riddled with convention, and lecturers often check this area very thoroughly.
- There are standard abbreviations for the journals. Don't make them up – ask the librarian. One of the most convenient publications for checking journal abbreviations is *Periodical Title Abbreviations*, Volumes 1–3, edited by L.G. Alkire, and published by Gale Research Company, Detroit, Michigan.

Examples of how to list the various sources

1 **Papers in journals:**

- Surname and initials of the author(s) (surname first, followed by the initials).
- The year of publication (in brackets).
- Title of the paper.
- The name of the journal (in italics or underlined and in its correctly abbreviated form. For instance, the journal abbreviation in the first example below (Bull. Inst. Math. App.) is the correct way to cite the Bulletin of the Institute of Mathematics and its Applications. The abbreviation in the second example is that for Scientific American.
- The volume number of the journal (underlined or in bold).
- The numbers of the pages on which the paper begins and ends. **Note:** the actual page from which your information is taken is not cited.

Examples:

Hart, V. G. (1982) The law of the Greek catapult. *Bull. Inst. Math. App.* **18**, 58–63

Soedel W. and Foley, V. (1979) Ancient catapults. *Sci. Am.* **240**, 150–160.

2 **Books:**

- Surname and initials of the author(s) (surname first, followed by the initials).
- The year of publication.
- Title of the book (underlined or in italics, and with the 'main' words (everything except articles, prepositions and conjunctions) capitalised.
- If there is a subtitle, it is separated from the main title by a colon (:).
- Title of series, if applicable.
- Volume number or number of volumes, if applicable.
- Edition, if other than the first.
- Publisher.
- Place of publication.
- Page numbers of the material quoted. **Note:** if you need to cite different parts of a book, it is acceptable to leave out the page numbers.

Examples:

Stroustrup, B. (1991) *The C++ Programming Language*. Second edition. Addison-Wesley, Reading, Massachusetts, pp 225–253.

Barrett, C.S. and Massalski, T.B. (1980) *Structure of Metals: Crystallographic Methods, Principles and Data*. Third edition. Pergamon Press, Oxford, pp 73–89.

3 **A chapter or article in a book edited by someone else: the 'In' citations:**

- Surname and initials of author(s).
- The year of publication.
- Title of chapter or article in quotation marks.
- Title of book, underlined or in italics. This is preceded by 'In:'.
- Volume number, if applicable.
- Editor(s) – preceded by 'Ed:' or 'Eds:'.
- Publisher.
- Place of publication
- The number of the pages on which the chapter begins and ends.

Examples:

Hall, J.E. (1992) 'Treatment and use of sewage sludge'. In: *The Treatment and Handling of Wastes*. Eds: A.D. Bradshaw, R. Southwood, and F. Warner. Chapman and Hall, London. pp 63–82.

Thomas, C.J.R. (1993) 'The polymerase chain reaction'. In: *Methods in Plant Biochemistry, Vol. 10: Molecular Biology.* Ed: J. Bryant, Academic Press, London. pp 117–140.

4 **Paper in the proceedings of a conference:**

- Author(s) / date / title of paper as for a journal paper (above), but in addition:
- State the number of the conference, its title theme, the place it was held and the date.

Bhattacharya, B., Egyd, P., and Toussaint, G.T. (1991) Computing the wingspan of a butterfly. *Proc Third Canadian Conference in Computational Geometry (Vancouver)*, Aug 6–10. pp 88–91.

5 **Student project:**

Cox, M.J.M. (1994) Improvement of a hang-glider's stall characteristics. Mechanical Engineering project, School of Engineering, The University of Middletown.

6 **Newspaper article:**

When the author is known:

Nicholson-Lord, D. (1995) Does work make you stupid? *Independent on Sunday*, 29 January, p 21.

When the author is unknown:

Northern Herald (1995) Rare bird ruffles experts' theories. 12 January, p 20.

7 **Thesis:**

Inman, M.E. (1994) Corrosion of carbon steel in geothermal systems. PhD thesis, The University of Middletown.

8 **Magazine article:**

Where the author is known, put the author first:
Schaer, C. (1993). Gene genius. *More*, December, 70-76

Where the author is not stated, put the name of the magazine first:

Consumer (1993) Shades of green. Number 322, p 16–19.

9 **Engineering standards**

Include both the title and the reference number.

ACI Committee 318, 1989. *Building Code Requirements for Reinforced Concrete and Commentary*, American Concrete Institute, Detroit.

10 **Government and legal documents:**

Cite the complete title.

World Health Organisation (WHO) (1977) *Manual of the Statistical Classification of Diseases, Injuries and Causes of Death: Based on the Recommendations of the Ninth Revision Conference, 1975, and Adopted by the Twenty-ninth World Health Assembly*, Vol. 1, WHO, Geneva.

This would be cited in the text as WHO (1977).

Department of the Environment (1988) *Integrated pollution control: a consultation paper*. DoE: London.

11 Personal communications:

Personal communications are not cited in the References section. If you have a number of them and to give them authenticity, you may like to have a separate section for them, citing the surnames, initials and affiliations of the people cited.

12 Lecture handouts:

If the writer's name is stated:

Seidel, R. (1996) Robotics. Lecture handout, Engineering and Society, The University of Middletown.

If the writer is unknown:

Wetlands (1996). Lecture handout, Conservation Ecology, The University of Middletown.

13 Laboratory manual:

Strain measurement (1996). Year Two Mechanical Engineering Laboratory Manual, The University of Middletown, 46–49.

14 Material from the Internet:

The Internet, particularly the Frequently Asked Questions sections, is increasingly being accessed by students for material relevant to their project work. It has to be remembered that these citations cannot be regarded as being as solidly based as those of the conventional sources, which are accessible via libraries and which will reliably exist over a long period of time.

There are not as yet any conventions as to how to cite this material. The following suggestions are only tentative:

- It may be wise not to include references to the Internet in the References section itself, but to follow with a separate section called **Internet sources**.
- The source could then be cited as follows:

Internet newsgroup 'comp. compression' (1995) Frequently Asked Questions Part 1, Subject [17]: *What is the state of fractal image compression?* Entry from P Mair <mair@zariski.harvard.edu>.

Such a source could alternatively be cited as a personal communication.

Examples of text and the corresponding References section

These examples show how to cite material in the text and the corresponding References section for the author-date system and the numbering system. The same text passage and citations are used for each one.

Example: Author-date system:

Notes:
Precise placing of reference, referring to the student project

Repeat of a previously-cited reference

Author mentioned in text

Three references in a series, placed in chronological order, separated by semi-colons

An 'et al.' reference – more than two authors

Precise placing of references in the text; one referring to the palintonon, and a different one to the onager

Text

The recent reconstruction of a trebuchet, the medieval siege engine, as a student project in engineering (O'Connor, 1994) has provided fascinating new insights into the mechanical efficiency of these hurling devices. Used in ancient times to hurl everything from rocks to plague-ridden carcasses of horses (O'Leary, 1994) and, in a modern four-storey-high reconstruction, dead pigs, Hillman cars and pianos (O'Connor, 1994), the trebuchet relied on the potential energy of a raised weight. Its mechanical efficiency has been compared unfavourably by Gordon (1978) with that of the palintonon, the Greek hurling device, which could hurl 40 kg stone spheres over 400 metres (Hacker, 1968; Marsden, 1969; Soedel and Foley, 1979). This device incorporated huge twisted skeins of tendon, a biomaterial that can be extended reversibly to strains of about 4% (Wainwright et al., 1976). The palintonon utilised the principle of stored elastic strain energy – the fact that when a material that has been deformed is unloaded it returns to its undeformed state due to the release of stored energy (Benham and Crawford, 1987). The motion of the palintonon (Hart, 1982) and that of its Roman equivalent, the onager (Hart and Lewis, 1986), has been analysed by use of the energy principle applied to the finite torsion of elastic cylinders.

Book. Note publisher, place of publication (Harlow) and relevant pages

Book

Chapter in book. The book is Volume 9 of a series called Technology and Culture. An 'In:' reference.

Paper in journal

Paper in journal

Book

Article in journal, no volume number

Editorial in journal

Article in magazine with volume number

More than two authors. An 'et al.' reference in text

References

Benham, P.P. and Crawford, R.J. (1987) *Mechanics of Engineering Materials*. Longman Scientific and Technical, Harlow, pp 66–68.

Gordon, J.E. (1978) *Structures or Why Things Don't Fall Down*. Penguin, Harmondsworth, pp 78–89.

Hacker, B.C. (1968) 'Greek catapults and catapult technology: science, technology and war in the ancient world.' In: *Technology and Culture*, **9**, No. 1, pp 34–50.

Hart, V.G. (1982) The law of the Greek catapult. *Bull. Inst. Math. Appl.*, **18**, 58–68.

Hart, V.G. and Lewis, M.J.T. (1986) Mechanics of the onager. *J. Eng. Math.*, **20**, 345–365.

Marsden, E.W. (1969) *Greek and Roman Artillery*. Clarendon Press, Oxford, pp 86–98.

O'Connor, L. (1994) Building a better trebuchet. *Mechanical Engineering*, January, 66–69.

O'Leary, J. (1994) Reversing the siege mentality. *Mechanical Engineering*, January, 4.

Soedel, W. and Foley, V. (1979) Ancient catapults. *Scientific American*, **240**, 150–160,

Wainwright, S.A., Biggs, W.D., Currey, J.D. and Gosline, J.M. (1976) *Mechanical Design in Organisms*. Princeton University Press, Princeton, pp 88–93.

Example: Sequential numbering system

Notes:
Another reference to Reference Number 1. Note that it is not assigned a new number

Author mentioned in text

Three references in a series, separated by commas

Precise placing of references in the text; one referring to the palintonon, and a different one to the onager

Text

The recent reconstruction of a trebuchet, the medieval siege engine, as a student project in engineering [1] has provided fascinating new insights into the mechanical efficiency of these ancient hurling devices. Used in ancient times to hurl everything from rocks to plague-ridden carcasses of horses [2] and, in a modern four-storey-high reconstruction, dead pigs, Hillman cars and pianos [1], the trebuchet relied on the potential energy of a raised weight. Its mechanical efficiency has been compared unfavourably by Gordon [3] with that of the palintonon, the Greek hurling device, which could hurl 40 kg stone spheres over 400 metres [4, 5, 6]. This device incorporated huge twisted skeins of tendon, a biomaterial that can be extended reversibly to strains of about 4% [7]. The palintonon utilised the principle of stored elastic strain energy – the fact that when a material that has been deformed is unloaded it returns to its undeformed state due to the release of stored energy [8]. The motion of the palintonon [9] and that of its Roman equivalent, the onager [10], has been analysed by use of the energy principle applied to the finite torsion of elastic cylinders.

Article in journal, no volume number

Editorial in magazine

Book. Note publisher, place of publication (Harmondsworth) and relevant pages

Article in magazine with volume number (240) in bold-face

Chapter in book. The book is Volume 9 of a series called Technology and Culture. An 'In:' reference

Book

Reference with more than two authors

Book

Paper in journal

Paper in journal

References

1 O'Connor, L. (1994) Building a better trebuchet. *Mechanical Engineering*, January, 66–69.

2 O'Leary, J. (1994) Reversing the siege mentality. *Mechanical Engineering*, January, 4.

3 Gordon, J.E. (1978) *Structures or Why Things Don't Fall Down*. Penguin, Harmondsworth, pp 78–89.

4 Soedel, W. and Foley, V. (1979) Ancient catapults. *Scientific American*, **240**, 150–160,

5 Hacker, B.C. (1968) 'Greek catapults and catapult technology: science, technology and war in the ancient world.' In: *Technology and Culture*, **9**, No. 1, pp 34–50.

6 Marsden, E.W. (1969) *Greek and Roman Artillery*. Clarendon Press, Oxford, pp 86–98.

7 Wainwright, S.A., Biggs, W.D., Currey, J.D. and Gosline, J.M. (1976) *Mechanical Design in Organisms*. Princeton University Press, Princeton, pp 88–93.

8 Benham, P.P. and Crawford, R.J. (1987) *Mechanics of Engineering Materials*. Longman Scientific and Technical, Harlow, pp 66–68.

9 Hart, V.G. (1982) The law of the Greek catapult. *Bull. Inst. Math. Appl.*, **18**, 58–68.

10 Hart, V.G. and Lewis, M.J.T. (1986) Mechanics of the onager. *J. Eng. Math.*, **20**, 345–365.

Using direct quotations

You may occasionally need to quote word-for-word from another source. This may be particularly so in essays where you are writing about contentious issues and feel that the exact words are relevant to your discussion. These quotations need to be enclosed in quotation marks. The conventions associated with this are:

1 Occasionally you may need to make very slight changes to a quotation so that it fits into your prose. For example, a capital letter may need to be changed to lower case, or you may think that you need to substitute a noun for a pronoun, or insert a noun, so that it makes more sense. These changes are indicated by square brackets [ ].

Examples:

Pratchett (1993) has noted that '[she] lived in the kind of poverty that was only available to the very rich, a poverty approached from the other side'.
(The original quote: 'Sybil Ramkin lived in ...'.)

Steven Jay Gould (1985) has stated that 'the history [of human races] is largely a tale of division – an account of barriers and ranks erected to maintain the power and hegemony of those on top'.
(The original quote: 'The history is largely ...'.)

2 If you leave out part of a quote because it is irrelevant to your document, use three dots to show the omission. It is very important that the sense of a quotation is not altered by the omission.

Example:

Inkster (1991) has said that 'the role of technological change … may be exaggerated but it may also be underestimated'.

Compiling a Bibliography

The conventions used for compiling a Bibliography are:

- Use the same method of writing out each item as you would for a References section (see pages 107–111).
- The items are listed in alphabetical order according to the first author's surname, or the title of the reference if the author is unknown.
- The list is not numbered.
- It is common practice to indent each line of a reference after the first.

Example:

Bibliography

Benham, P.P. and Crawford, R.J. (1987) *Mechanics of Engineering Materials.* Longman Scientific and Technical, Harlow, pp 66–68.

Gordon, J.E. (1978) *Structures or Why Things Don't Fall Down.* Penguin, Harmondsworth, pp 78–89.

Hacker, B.C. (1968) 'Greek catapults and catapult technology: science technology and war in the ancient world.' In: *Technology and Culture,* **9**, No. 1, pp 34–50.

Hart, V.G. (1982) The law of the Greek catapult. *Bull. Inst. Math. Appl.,* **18**, 58–68.

Hart, V.G. and Lewis, M.J.T. (1986) Mechanics of the onager. *J. Eng. Math.,* **20**, 345–365.

Marsden, E.W. (1969) *Greek and Roman Artillery.* Clarendon Press, Oxford, pp 86–98.

O'Connor, L. (1994) Building a better trebuchet. *Mechanical Engineering,* January, 66–69.

O'Leary, J. (1994) Reversing the siege mentality. *Mechanical Engineering,* January, 4.

Soedel, W. and Foley, V. (1979) Ancient catapults. *Scientific American,* **240**, 150–160,

Wainwright, S.A., Biggs, W.D., Currey, J.D. and Gosline, J.M. (1976) *Mechanical Design in Organisms.* Princeton University Press, Princeton, pp 88–93.

Common mistakes

- Citing a reference in the text and leaving it out of the References list.
- Citing a reference in the References list and making no mention of it in the text.

These two mistakes tend to be regarded as unforgivable by most lecturers.

- Using non-standard abbreviations for a journal.
- Giving insufficient details in the References section: in particular, omitting the publisher and place of publication of a book, omitting the date.
- Not inverting the author and the initials: for example, instead of the correct form of Smith, A.N., putting A.N. Smith.

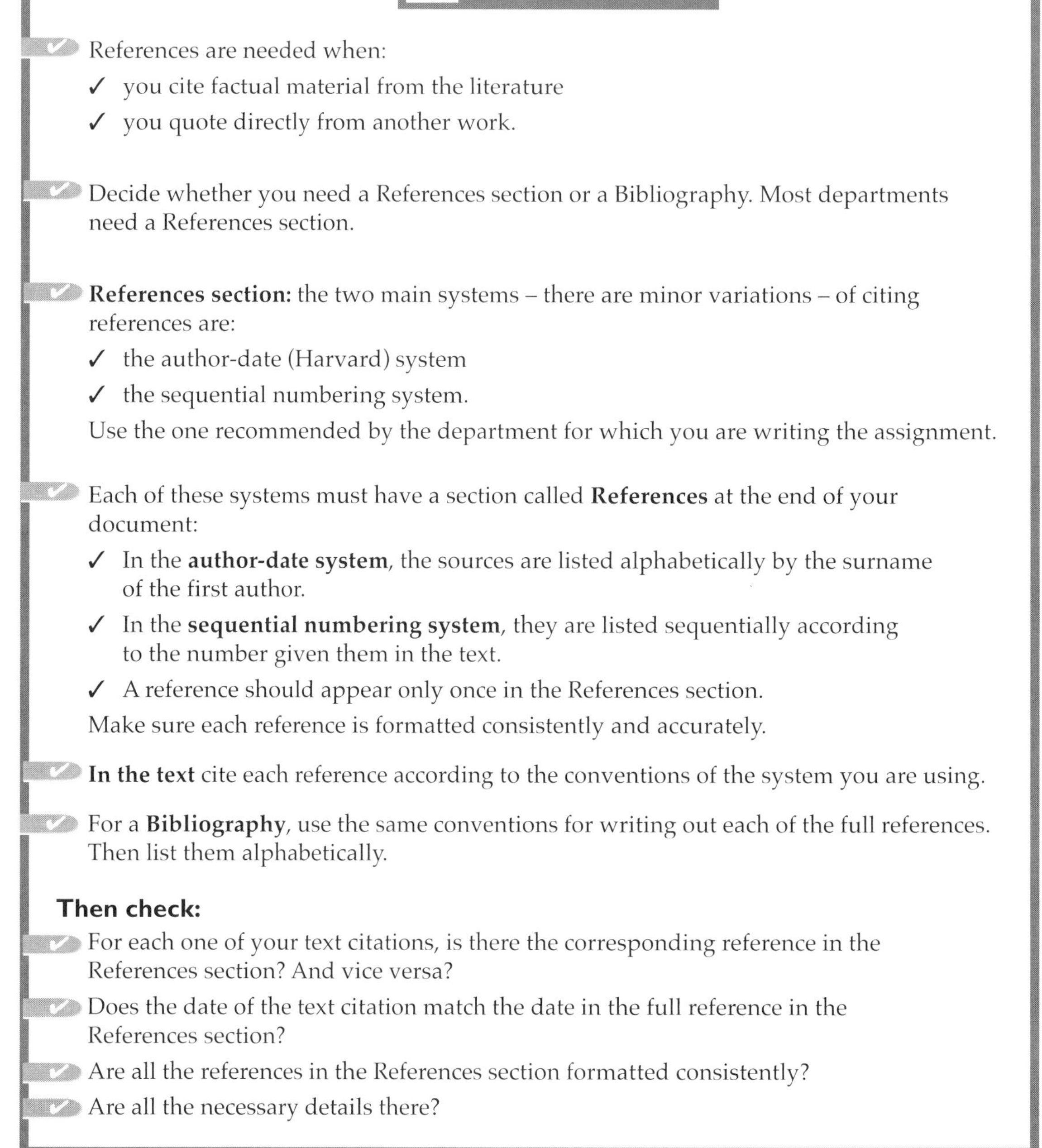

Checklist: References

- References are needed when:
  - ✓ you cite factual material from the literature
  - ✓ you quote directly from another work.

- Decide whether you need a References section or a Bibliography. Most departments need a References section.

- **References section:** the two main systems – there are minor variations – of citing references are:
  - ✓ the author-date (Harvard) system
  - ✓ the sequential numbering system.

  Use the one recommended by the department for which you are writing the assignment.

- Each of these systems must have a section called **References** at the end of your document:
  - ✓ In the **author-date system**, the sources are listed alphabetically by the surname of the first author.
  - ✓ In the **sequential numbering system**, they are listed sequentially according to the number given them in the text.
  - ✓ A reference should appear only once in the References section.

  Make sure each reference is formatted consistently and accurately.

- **In the text** cite each reference according to the conventions of the system you are using.

- For a **Bibliography**, use the same conventions for writing out each of the full references. Then list them alphabetically.

Then check:

- For each one of your text citations, is there the corresponding reference in the References section? And vice versa?
- Does the date of the text citation match the date in the full reference in the References section?
- Are all the references in the References section formatted consistently?
- Are all the necessary details there?

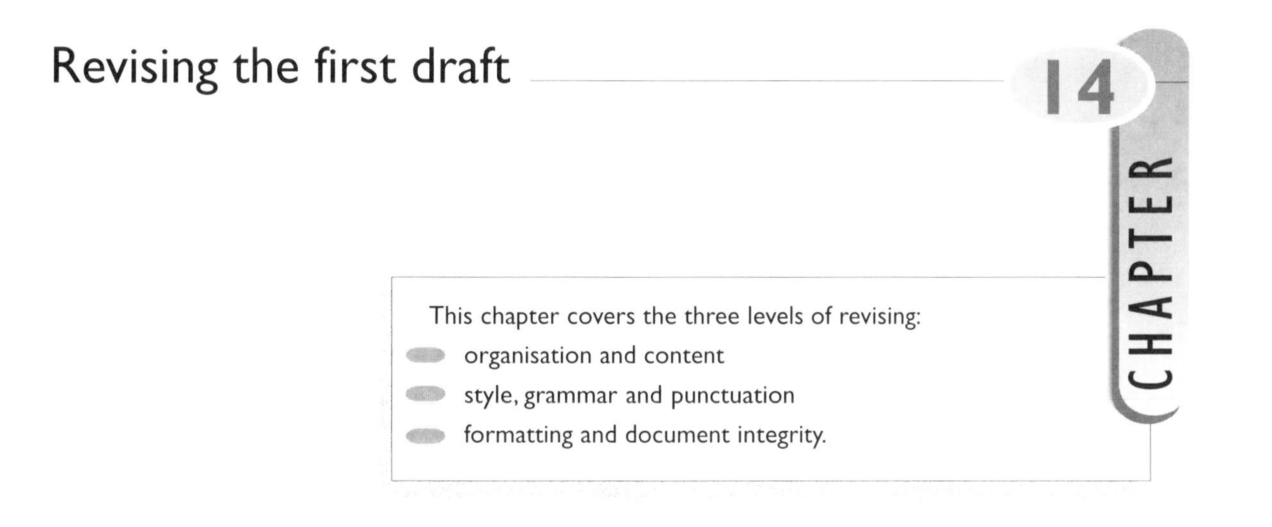

Revising the first draft

This chapter covers the three levels of revising:

- organisation and content
- style, grammar and punctuation
- formatting and document integrity.

The main aspects of the revising process

1 Allow plenty of time for rewriting

The final stages of the writing process – revising, retyping or rewriting, inserting illustrations etc. – always take much, *much* longer than expected. Allow ample time for them in your planning schedule, otherwise you'll find yourself having to submit your first draft.

2 Stand back from it

After writing something, most people become blind to what they have written. They find it difficult to review it objectively, and can see only minor typographical errors. There are two ways of overcoming this:

- **Ask someone else to read it.** If you are not in a position to do this, then:
- **Stand back from it for as long as possible.** Most undergraduates don't have the time; however try to put it aside for as long as you can. Even half an hour helps; a day or more is much better. Mistakes of organisation and style then leap out at you.

3 Use a printout. Don't try to do it off the monitor

If you have word-processed your assignment, don't do your editing off the monitor. It is essential to use a printout; mistakes are far more obvious.

4 Three-level revising

You will greatly increase your efficiency and save time if you break it down into three separate steps:

Level One: **organisation and content only**
Level Two: **style, grammar and punctuation**
Level Three: **formatting and document integrity**

Level One: organisation and content only

At this first stage it is important to:

- **Concentrate on just the organisation and content.** Examine whether the **structure** is right.
- **Actively ignore style, typographical errors and punctuation.**

When you are revising a draft for the first time, you will find that the easiest things to notice are errors of style, spelling and punctuation. If you allow yourself to get immersed in this fine detail, you won't be able to pick out errors of organisation, and you will have lost any advantage gained from standing back from it for a while.

Level One checklist: organisation and content

- **Examine your assignment critically in an *overall* way. Look only for errors of organisation.**
  - **a** Have you followed your plan?
  - **b** Is the structure logical?
  - **c** Should this section come before that one, that paragraph before this one?
  - **d** Is the information coherent?
- **While you are concentrating on the organisation, rapidly make margin marks to show errors of style. Don't allow yourself to be drawn into correcting them.** Come back to them at the next stage. Be aware that you have to discipline yourself to ignore detail at this stage. It's quite difficult.

Level Two: style, grammar and punctuation

At this stage, go back and polish up detail of style.

Level Two checklist: style, grammar, punctuation

(see Chapter 15 **Problems of style: how to correct them**, pages 125–158)

	Page number
Writing for your audience	
• Are you writing for your reader, not for yourself?	37
Paragraphs	
• Are the breaks between the paragraphs logical?	135
Sentences	
• Are you writing in 'real' sentences?	126
• Are your sentences too long?	135
• Are they bitty? Do they need linking?	133
Verbs	
• Are you using the distorted passive or other lifeless verbs?	135
• Are you using the right tense?	139
• Is there subject-verb agreement?	139
Words	
• Have you spell-checked it?	140
• Are you using pompous words where short ones would be better?	141
• Are you using jargon or clichés?	142
• Are you using *I* or *We* too often?	143
• Is your language gender-neutral?	143
• Are there colloquialisms or contractions?	145
• Are you using the right word? (*Affect/effect, led/lead, lose/loose, etc.*)	146
• Are you writing numbers correctly? (*Ten or 10?*)	154
Punctuation	
• Are your commas and full stops effective?	155
• Are the apostrophes used correctly?	151

Level Three: formatting and document integrity

Formatting

An assignment should be visually strong, not dauntingly dull-looking. But make sure that each formatting decision you make is for a good reason. If you just play with it, you'll end up with a document that looks disordered.

If you are word-processing your assignment, you have many options:

1. **For emphasis: boldface type, Larger font** or **Different font** for headings. But use these with discretion. Too many different fonts can look messy.

2. **Paragraphs.** The two methods for formatting the first line of a paragraph are:

Formatting	**Text**
1 Leave one line between each paragraph (press Enter twice). No indentation.	A point about half way along a hummingbird wing moves only about 5 chord lengths in each cycle of wing movements. Hummingbirds are not the only flying animals that get more lift than steady-state aerodynamics predicts.
2 No space, but indent the first line by no less than 5 characters.	A point about half way along a hummingbird wing moves only about 5 chord lengths in each cycle of wing movements. Hummingbirds are not the only flying animals that get more lift than steady-state aerodynamics predicts.
Common mistake: no space, and an indentation of only one or two characters. This makes it very difficult to distinguish one paragraph from the next.	**What not to do:** A point about half way along a hummingbird wing moves only about 5 chord lengths in each cycle of wing movements. Hummingbirds are not the only flying animals that get more lift than steady-state aerodynamics predicts.

3. **Justified right-hand margins.**

4. **Reduce the packing density.** Information is more readily absorbed if it's not too dense on the page:
   - **Wide margins.**
   - **Bullet points in the text of reports** (page 68). This is a powerful method for listing within the text.
   - **Indented left-hand margins,** for instance when using bullet points or quotations. However, avoid indenting so many times that the text is squeezed into the right-hand side of the page.

5 **The amount of space between lines.** But be careful: if, in the main text, the line spacing is too wide (e.g. double spacing) and/or the font is too large (larger than 12 point), the marker's overwhelming impression is that you are trying to hide lack of content.

6 **Effective page breaks.** Avoid the following bad breaks:
- a heading at the bottom of the page (there should be at least two lines of text following a heading)
- a short line (a widow) at the top of the page
- a table that is cut in two by a page break
- a page that ends with a hyphenated word.

Document integrity

The changes made during editing may cause a mismatch between parts of a document.

Common problems are:

1 Discrepancies in:
- the numbering of section headings
- referring to figures in the text
- referencing
- the Contents page and text page numbers.

2 Missing figures, tables or sections of text.

Level Three checklist: document integrity

Illustrations (see Chapter 12 ***Illustrations****, pages 89–102)*

Is each illustration:	*Page*
• numbered?	99–100
• titled?	99-100
• adequately labelled?	93
• correctly referred to in the text?	100

References (see Chapter 13 ***References****, pages 103–116)*

- For each text citation, is there the corresponding reference in the References section? And vice versa?
- Does the date of the text citation match the date in the full reference in the References section?
- Are all the references in the References section formatted consistently?
- Are all the necessary details there?

Text	***Page***
Are the headings and subheadings numbered consistently?	53

Contents page (see pages 52-55)

- Does the wording of headings match up with the text headings?
- Is the numbering of each heading and its subheadings consistent?
- Is the formatting (indenting) of the Contents page consistent?
- Are the page numbers correct?

Writing style

Problems of style: how to correct common mistakes

CHAPTER 15

This chapter does not try to give comprehensive guidelines on stylistic elegance. Instead, using simple terms, it tries to address problems often met by science and engineering undergraduates. The Further Reading section gives suggestions about where to find more information on style in writing.

Style is as important as content. Many students believe that if they have the right content to an assignment, they will get a high grade. But if your assignment is difficult to read – if it is badly or boringly worded – the grade will drop, even if the content is good.

You need to show your marker that you can write coherently and articulately. But many science and engineering students freely acknowledge that they feel very unprepared to be able to do this. This section looks at some of the specific mistakes in style often made in undergraduate science writing.

Use Appendix 1 if you need to brush up on:

- the meanings of words such as *noun*, *verb*, etc.
- tenses of the verb.

Avoid transcribing out of books: have the confidence to put it in your own words

Be confident in your ability to express something in your own words. This is essential if you want to avoid one of the commonest aspects of undergraduate writing – copying reams of material, sometimes word-for-word, sometimes slightly paraphrased, out of books on the reading list. It is lack of confidence and, perhaps, laziness that makes people do this. But the result is obvious to the marker – noticeable pieces of 'textbook-ese', strung together with marked discontinuities of style, making for a disjointed piece of writing. Moreover it is plagiarism – literary theft – and as such is regarded very seriously (see page 103).

To avoid copying out chunks of material, we need to be confident of our abilities to think and write for ourselves.

First, we have to know what are the features of a good writing style. The rest of this chapter, using simple terminology, sets out to do the following:

- it highlights specific errors common in the writing of science and engineering students
- it gives practical suggestions about how to correct them.

Writing good sentences

Rambling sentences: be clear, straightforward and concise

The first aim in writing is to be readily understood. This needs clarity. It doesn't mean that the content is simplistic; it just means that it is expressed in the most straightforward way possible.

Sentences made up of a string of components joined with *and* or *but*.

Many writers find it easier to write long, complex sentences than straightforward ones, often by stringing together a number of ideas using words like *and* and *but* to join them up. In general, short sentences are easier to understand than long ones. So disentangle these long sentences into shorter ones.

Example:

Original long sentence made up of several components joined up together by *and* or *but*.
Companies involved in the pollution had two choices, either clean up or pay a small fee to the government and, as a result, pollution continued as it was cheaper for them to pay the fine than to clean up, but this is changing and companies are now more aware of the problem and they include decontamination in their cost estimates, but the system is far from perfect.

Sentence rewritten by getting rid of the linking *and's* and *but's*, and making five sentences.
Companies involved in the pollution had two choices: either clean up or pay a small fee to the government. As a result, pollution continued as it was cheaper for them to pay the fine than to clean up. This is changing. Companies are now more aware of the problem and they include decontamination in their cost estimates. However, the system is far from perfect.

Putting it right: Rambling sentences

Examine a sentence to see if it's made up of a number of elements joined up by words such as *and*, *but* etc. Chop it up into smaller sentences.

Use real sentences: avoid sentence fragments

How do you recognise a real sentence?

Writing sentences that aren't sentences – we'll call them sentence fragments – is one of the main characteristics of student writing, and consequently one of the main grumbles of staff markers.

Definition: a sentence

A sentence has a full-stop at the end of it, and usually contains one main idea. It must have a subject and a finite verb.

There are two main ways in which people write sentence fragments:

1 Not having a finite verb.

2 Using *Which* at the beginning of a sentence when it's not a question.

Not having a finite verb. What is it?

Definition: a finite verb is one that is limited by person and number

In the sentence:

The waves are eroding the shore.

the verb *are eroding*, is limited – or defined – by the fact that it's the waves that is the subject of the sentence. It's limited by being third person plural. The words:

Eroding the shore.

cannot stand alone as a sentence (i.e. something between full stops) because the verb is not finite – it is not limited or defined because it is not referring to anything. It doesn't sound complete.

We will clarify this more by looking at specific examples:

Most people can recognise a verb. It's the *doing* word (see Appendix 1 for an explanation of the parts of speech).

Examples:

The wind blew
The waves pounded.
We climbed the mountain.

There is another way we can put these:
The wind was blowing.
The waves were pounding.
We were climbing the mountain.

In this case the verb is made up of two words *was/were* and a word ending in *-ing*. However, the word ending in *-ing* can also be used another way, by itself, without *was/were/will be* in front of it. And this is where the problem starts to arise. We can use *...ing* in another part of the sentence, the subclause.

The wind was blowing, making the trees sway.
The waves were pounding, causing some damage to the sea wall.
We were climbing the mountain, stopping sometimes for a rest.
Main clause with a finite verb — Sub-clause without a finite verb

The *main clause* in these sentences is the clause that can stand alone and make sense. The *subclause* cannot make sense by itself.

A frequent problem in writing results from trying to make that other part of the sentence – the sub-clause – into a sentence by itself, by putting a full stop where the comma is. This forms a *sentence fragment.*

Examples

The eruption caused a huge loss of farm animals and crops. This being devastating for the farmers.

Increasing agriculture will cause an increase in global warming. The reason being that ruminants and paddy fields produce methane.

Table 15.1

Original incorrect version	**Corrected versions. Corrected using:** • **Method 1 (a comma) or** • **Method 2 (using a finite verb in the second sentence)**
Increasing agriculture will cause an increase in global warming. The reason being that ruminants and paddy fields produce methane.	**Method 1** Increasing agriculture will cause an increase in global warming, the reason being that ruminants and paddy fields produce methane. **Method 2** Increasing agriculture will cause an increase in global warming. The reason is that ruminants and paddy fields produce methane.
The eruption caused a huge loss of farm animals and crops. This being devastating for the farmers.	**Method 1** The eruption caused a huge loss of farm animals and crops, this being devastating for the farmers. **Method 2** The eruption caused a huge loss of farm animals and crops. This was devastating for the farmers.
Shops and buildings were reduced to ruins and the streets covered with damaged cars and rubble. The time being frozen forever on the hands of the band rotunda at 10.47 am.	**Method 1** Shops and buildings were reduced to ruins and the streets covered with damaged cars and rubble, the time being frozen forever on the hands of the band rotunda at 10.47 am. **Method 2** Shops and buildings were reduced to ruins and the streets covered with damaged cars and rubble. The time was frozen forever on the hands of the band rotunda at 10.47 am.
There are a number of strategies that countries can take. For example, promoting non-wood fuel sources, paper recycling and pricing forest products more efficiently.	**Method 1** There are a number of strategies that countries can take, for example, promoting non-wood fuel sources, paper recycling and pricing forest products more efficiently. **Method 2** There are a number of strategies that countries can take, For example, they can promote non-wood fuel sources, recycle paper and price forest products more efficiently.

Table 15.1 shows examples taken from student reports. Look first of all at Column 1 – the original incorrect versions. In each case the first sentence is correct. The second 'sentence' has these characteristics:

- there is a verb form ending in *-ing* (very often *being*). It is generally found near the start of the 'sentence'
- there is no finite verb.

Putting it right: sentence fragments where ***-ing*** *has been used*

How to recognise it:

- Do you have these two things together? (see column 1, Table 15.1)
  a Somewhere near the beginning of a 'sentence', a word ending in *-ing? Being* is very common.

  and

  b Is there no finite verb in the 'sentence', or in the first clause of that 'sentence'? Disregard infinitives i.e. to + verb.
- Then there is a good chance that you have a sentence fragment.

The favourite: Using *This being....* or *The reason being....* to start a sentence.

Two ways to put it right: (see column 2, Table 15.1)

Method 1
In most cases, it is enough to put a comma instead of a full-stop. Look at Method 1 in the corrections of the examples in Table 15.1.

Method 2
The passage often reads better if you make two sentences, and use a finite verb in the second one instead of the word ending with *-ing*. See Method 2 in the examples in the table.

If you're still puzzled, try saying it to yourself. Use instinct. A sentence fragment will generally sound odd. If you say *'This being devastating for farmers.'* completely in isolation, it sounds incomplete. On the other hand, *'This was devastating for farmers.'* sounds complete. Anything between two full-stops should sound complete in itself.

Using *Which* at the beginning of a sentence (when it's not a question)

Another way of making a sentence fragment is by starting a 'sentence' with ***Which***, when you're not asking a question. It's very common to see it in commercial material, for example:

> All of these plans have been designed with you in mind. Which is why you'll find one that's just right for you.

However, this is unacceptable in technical writing.

Putting it right: sentence fragments starting with **Which**

How to recognise it:

When you have *Which* as the first or second word in a 'sentence', and it's not a question.

Two ways of putting it right (see Column 2 in Table 15.2)

Method 1
Put a comma instead of a full-stop.

Method 2
If putting a comma makes the sentence too long, rewrite the second part. You can generally start the second sentence with *This is....*

Column 1 in Table 15.2 shows an example of a very common mistake in student assignments:

Table 15.2

Original incorrect version	Corrected versions. Corrected using: • Method 1 (a comma) or • Method 2 (starting another sentence using This ...)
When the engine was run on petrol the carbon dioxide emissions were higher. Which was an indication of improved mixing and less cylinder-to-cylinder variation.	**Method 1** When the engine was run on petrol the carbon dioxide emissions were higher, which was an indication of improved mixing and less cylinder-to-cylinder variation. **Method 2** When the engine was run on petrol the carbon dioxide emissions were higher. This was an indication of improved mixing and less cylinder-to-cylinder variation.

Summary: Sentence fragments

Ways of recognising sentence fragments:

- Look for words ending in *-ing* just after a full-stop.

- Look for *Which* right at the beginning of a sentence, when it's not a question.
- Say it out loud. Use your instinct. If it sounds incomplete, it probably needs adjustment.

Comma splices

This is rather like the reverse of the sentence fragment problem. People often link sentences together using commas, instead of full-stops. The result is a series of 'sentences' complete in themselves, but strung together with commas.

There are ways of joining sentences together using small words called conjunctions (linking words, see Appendix 1). The most common conjunctions are *and, but, so, because, as, while* and *since*. The problem occurs when two sentences are joined up without using a linking word. This happens when the second sentence tends to follow logically from the first, so people just use a comma.

Example:

Wrong The weather was good, we went to the beach.
First clause Second clause

Correct The weather was good so we went to the beach.
First clause Conjunction Second clause

Each of the two clauses in the wrong version is complete in itself, yet they are joined with only a comma. Such comma splices are very common in student writing, and are a source of irritation to markers. It pays to correct them.

Table 15.3 shows examples of comma splices from student writing. Column 1 shows the original incorrect examples. Column 2 shows corrected versions.

Table 15.3

Original incorrect version	**Corrected versions. Corrected using:** • **Method 1 (a full-stop) or** • **Method 2 (a linking word) or** • **Method 3 (a semi-colon)** **All three methods have been used in each case. However, some of the solutions are more elegant than others. You have to choose what sounds best.**
There will be a number of Internet developments in the next few years, one of these is computer video.	**Method 1** There will be a number of Internet developments in the next few years. One of these is computer video. **Method 2** There will be a number of Internet developments in the next few years, and one of these is computer video. **Method 3** There will be a number of Internet developments in the next few years; one of these is computer video.

As a result of an earthquake the sea floor may suddenly rise or fall, this will create a tsunami.	**Method 1** As a result of an earthquake the sea floor may suddenly rise or fall. This will create a tsunami. **Method 2** As a result of an earthquake the sea floor may suddenly rise or fall, and this will create a tsunami. **Method 3** As a result of an earthquake the sea floor may suddenly rise or fall; this will create a tsunami.
Low soil pH can damage plant roots, they become shorter and more brittle.	**Method 1** Low soil pH can damage roots. They become shorter and more brittle. **Method 2** Low soil pH can damage roots, so they become shorter and more brittle. **Method 3** Low soil pH can damage roots; they become shorter and more brittle.
In December 1952 a deadly smog hung over London for almost a week, it killed about 4000 people.	**Method 1** In December 1952 a deadly smog hung over London for almost a week. It killed about 4000 people. **Method 2** In December 1952 a deadly smog hung over London for almost a week, and it killed about 4000 people. **Method 3** In December 1952 a deadly smog hung over London for almost a week; it killed about 4000 people.
Batteries can be used together with solar cells, however they have a much shorter service life.	**Method 1** Batteries can be used together with solar cells. However they have a much shorter service life. **Method 2** Batteries can be used together with solar cells but, however, they have a much shorter service life. **Method 3** Batteries can be used together with solar cells; however they have a much shorter service life.

Putting it right: the comma splice

Recognising it

- You probably have a comma splice if, after a comma:
  - **a** you think that the rest of the sentence may be a complete sentence in itself

  and
  - **b** there's no linking word (*and, but, so, since, because*) immediately after the comma
- Very common: when there is a comma immediately before *however* (see the last example in Table 15.3).

Three ways of correcting it

(Use whichever sounds better in your particular case)

Method 1

Putting a full-stop instead of a comma. This will always work and is your safest bet.

The weather was good. We went to the beach.

Method 2

Using a conjunction, or linking word. Try it; it won't always work.

The weather was good so we went to the beach.

Method 3
Using a semi-colon. This will always work and is an elegant way of doing it.
The weather was good; we went to the beach.

Avoid bitty sentences

Most people know that long, convoluted sentences are bad style. However, many people go to the other extreme and write bitty sentences. These come in streams and the ideas are often unlinked, making them difficult to read.

By all means, keep your sentences short. This makes for easier digestibility. But streams of bitty sentences need to be linked.

Example:

Wind is a sustainable energy source. Energy costs will not rise. They may come down if the costs of equipment get lower. There has been rapid development in wind turbine generators. Most present wind turbine generators produce around 500 kW. There are larger machines of 750–1000 kW machines. These are being used for research and development. There is other current research on energy capture. Turbulent winds are also being looked at. Researchers are trying to improve manufacturing methods, too.

Improved by linking the sentences:

Wind is a sustainable energy source. Therefore energy costs will not rise and may even come down if the costs of equipment get lower. Recently there has been rapid development in wind turbine generators. Although most present wind turbine generators produce around 500 kW, larger machines of 750–1000 kW machines are being used for research and development. There is other current research on energy capture, turbulent winds and the improvement of manufacturing methods.

Putting it right: Using linking words to avoid bitty, unlinked sentences

This is a list of words that you can use which, when used at the beginning of a sentence, link that sentence to the previous one.

For expressing a cause
Accordingly
Consequently
For this reason
Hence
Therefore
Thus

To emphasise something
Above all
Certainly
Clearly
Indeed
In fact
In short
Obviously
Of course

To express intention
For this purpose
To do this
To this end
With this in mind

To amplify
Again
Also
Apparently
Besides
Equally important
Finally
First, Second etc.
Furthermore
In addition
Moreover

To express location
Beyond
Here
There
Nearby
Opposite
Overlying (underlying)
To the right (left)

To concede something
At any rate
At least

Giving an example
For example
For instance
To illustrate

Time
Afterwards
At the same time
Before
Earlier
In the meantime
Sometimes
Later
Next
Preceding this
Recently
Simultaneously
Soon
Until

To qualify something
Although
Even though

For detail
Especially
In particular
Namely
Specifically
To enumerate

Interpreting something
Fortunately
(unfortunately)
Interestingly
Significantly
Surprisingly
(unsurprisingly)

Closing an item
In conclusion
To summarise

To generalise
On the whole
In general
Generally speaking
Broadly
Broadly speaking

Dissimilarity, contrast	**Similarity**
However	Likewise
In contrast	Similarly
Nevertheless	
On the contrary	
On the other hand	

How long should a sentence be?

Short sentences are more digestible. You can also get into less trouble with their construction. Modern writing tends to favour sentences of roughly 20–25 words. However, don't treat this as an absolute. Your readers will be bored if you deal them equal-length sentences one after another. The occasional longer sentence, if it is well-crafted and not overloaded with ideas, will make your writing more interesting.

Variety is important in sentence length. Aim for an *average* figure of 20–25 words per sentence, but oscillate around the mean.

How long should a paragraph be?

As with sentences, varying the length of paragraphs is another way of avoiding boring your reader. Avoid very long paragraphs. The psychological effect of black, uninterrupted text is extreme. People don't even want to start reading something that is so daunting.

Many students find effective paragraphing tricky; knowing that long paragraphs are bad, many seem to decide quite arbitrarily on where paragraph breaks should be placed. The result is an incoherent text.

It is difficult to give guidelines on how to paragraph effectively, but as a general rule it can be said:

- One main idea per sentence.
- One theme per paragraph. If there is a natural break in what you are writing, start a new one.
- The first sentence of a paragraph – *the topic sentence* – should introduce the theme of each paragraph.

Using verbs correctly

Vivid language: active, passive and distorted passive

Vivid language is not something that most people associate with technical writing. Yet if we feed people dull, impersonal prose we bore them. A lot of technical writing is dull, and much of the problem is to do with the way we use verbs.

Active versus passive voice

Many writing handbooks and word-processor grammar checkers

tell us to use the active voice of the verb, not the passive. This is not useful advice – most scientists and engineers have no idea what the active and the passive voices of the verb are.

We will now ask

- What is meant by the active and passive voices?
- Is using the passive voice bad?
- What happens when we distort the passive voice and make *really* impenetrable sentences?

What is the active voice of a verb?

Recognising the verb in a sentence

We are usually taught in school that a verb is the 'doing' word of a sentence. The following sentence has a subject, or actor (*acid-etching*), a verb (*removed*) and an object, or receiver (*rust*).

Acid-etching removed the rust. **Active voice of the verb**
Actor verb receiver

This sentence is in the **active voice** because the order of the flow is: Actor, verb, receiver.

Now turn this sentence around. We have:

The rust was removed by acid-etching. **Passive voice of the verb**
Receiver verb actor

When the order of flow is Receiver, verb, actor, the sentence is in the **passive voice**.

What happens when an active sentence is turned around into the passive voice?

- The order of the flow is reversed.
- The number of words in the verb increases – *removed* becomes *was removed* – as a result of adding forms of the verb *to be*.
- An extra word is needed (*by*).
- The emphasis has changed. In the active sentence the emphasis was on *acid-etching*; in the passive form *rust* is emphasised.

Is using the passive voice bad?

No. In technical writing we would say, quite naturally, *The pH was adjusted (passive)*, implying *The pH was adjusted by me*. We don't write *I adjusted the pH (active)*. The passive voice is not intrinsically bad, in spite of what many writing textbooks and grammar checkers tell us. Technical writing often needs the passive voice, otherwise we would be writing *we* or *I* all the time.

Sometimes, however, we must actively choose which voice of the verb to use. For instance, if we were writing a paragraph about bees and their relationship with pollen, we would write *Bees carry pollen (active)*. If the paragraph were about pollen, we'd write *Pollen is carried by bees (passive).* Each of these is completely acceptable; it depends on which emphasis we need.

Taking the passive voice one step further: the distorted passive

What is bad is what we as technical writers can do to the passive voice when we take it one step further. We distort it. We take the verb and hide it in a sort of a noun. This construction could be called the **Distorted Passive** voice.

Let's consider the progression in these sentences:

Acid-etching removed the rust	**Active voice**	Acceptable

Turn this around and it becomes:

The rust was removed by acid-etching	**Passive voice**	Acceptable

Hide the verb 'was removed' in a sort of a noun, and it becomes:

Removal of the rust was ...... by acid-etching Hidden verb a missing verb	**Distorted passive**	Tedious, pompous

Ask someone to insert the missing verb and the suggestions are always the same. The favourites are: **achieved, accomplished, carried out, performed, undertaken, effected, done.**
Now we've lost the skeleton of the sentence. We've gone from *Acid-etching removed*, or *The rust was removed* – both of which are fine - to *Removal was carried out*, or *Removal was achieved*, which sounds pompous.

This distortion is one of the commonest ways of writing tedious, impenetrable prose in science; we often go one stage further on from the passive than we need.

The ohmmeter measured the resistance	**Active voice**	Acceptable
Resistance was measured by the ohmmeter	**Passive voice**	Acceptable
Measurement of the resistance was carried out by the ohmmeter	**Distorted passive**	Tedious, pompous

I measured the leaf area daily	**Active voice**	Not generally acceptable
The leaf area was measured daily	**Passive voice**	The acceptable style for science writing
Daily measurement of leaf area was carried out.	**Distorted passive**	Unnecessary distortion

We are so used to seeing the distorted passive in professional writing that the absurdity of the construction is only obvious when it's seen in an everyday context:

Cinderella dropped the glass slipper	**Active**	Acceptable
The glass slipper was dropped by Cinderella	**Passive**	Acceptable
Dropping of the glass slipper was carried out by Cinderella	**Distorted passive**	Absurd

Putting it right: how do you rewrite the distorted passive?

- If you find yourself using *achieved, accomplished, carried out, performed, undertaken, effected,* you are very likely to be in the distorted passive. So keep these words in mind as danger signals.
- Find the hidden verb earlier in the sentence. It will probably be in a word ending with *...ing. ...tion* or *...ment.*
- Use it to rewrite the sentence, using either a simple passive construction or the active.

Avoiding the distorted passive is one of the most powerful tools for vivid writing in science.

Other lifeless verbs

Lifeless verbs halt the movement of a sentence. The worst offenders are *exist, occur,* and various forms of the verb *to be.*

Original lifeless version	Rewritten version
Increasing temperature occurred.	The temperature increased.
The purpose of this report is to describe the different stages of waste-water treatment.	This report describes the different stages of waste-water treatment.

Excessive use of nouns instead of verbs

Some lifeless verbs can mutate into nouns, and the pace slows down:

Indicates becomes *is an indication of*
Suppose becomes *make the supposition*

Original lifeless version with verb mutated to noun	Rewritten using verb
The colour of the outfall was an indication of severe pollution.	The colour of the outfall indicated (or showed) severe pollution.
We may therefore make the supposition that ...	We may therefore suppose that ...

The right tense of the verb

(see Appendix 1 for an explanation of tense).
Decisions about the proper use of tense can be muddling.

Here are guidelines for deciding which tense to use in reports:

1. Do not wander between present and past tenses. This is common.
2. Use the past tense when you are describing your own findings and procedures, and other people's results. This is the simplest way to keep your tenses consistent.
3. Use the present tense for describing illustrations:

Example:

Figure 10 shows the effect of temperature on the solubility of the salt in water.

You will probably also occasionally need the present tense in the Theory and the Discussion.

Examples:

Steel expands when heated. (Present tense – You are stating a known and accepted fact).

The steel expanded when heated. (Past tense – You are describing the results of your experiment).

Putting it right: verb tense

Don't mix up your tenses. Use past tense for just about everything except for describing illustrations, and occasionally in the Theory and the Discussion.

Subject/verb agreement

Make sure that the subject (actor) of your sentence agrees with the verb.

Original incorrect version	**Corrected version**
Mazda are the only company that has persevered with the rotary engine concept.	Mazda *is* the only company that has persevered with the rotary engine concept.
The greatest loss of lives as a result of a volcanic eruption have occurred through pyroclastic flows and tsunamis.	The greatest loss of lives as a result of a volcanic eruption *has* occurred through pyroclastic flows and tsunamis. **Note:** the verb is referring to *loss* (singular) not *lives* (plural)

Getting the words right

Check your spelling

Some people don't notice when words are spelled wrongly. Others not only notice; they find that it interrupts the flow of the sentence. This has a negative effect on the markers. Be aware that many lecturers fall into this second category, and that you need to check your work. Never underestimate the effect your bad spelling has on the quality of a piece of writing.

There are three ways of ensuring that your assignments don't contain spelling mistakes.

1 Use the spell-checker on the word-processor

- **Even if you are hurrying, your assignment should still be run through a spell-checker.** Do it; don't avoid it; it makes a cosmological difference to your marks.

  It is utterly amazing how many people submit word-processed assignments full of mistakes that a spell-check would have sorted out in only a few minutes.

- Even when you have spell-checked it, you can never assume that a spell-checked assignment is error-free. You have to proof-read it afterwards. A spell-checker will pass words that you may not have intended – *it* instead of *is*, *an* instead of *on*. You can end up with garbage, or sources of mirth of the sort found in an engineering assignment, where '300 revolutions per minute' was written as '300 revelations per minute'.

Here is a constructed example of the sort of garbage that can arise from a spell-checked passage that wasn't proof-read afterwards. Each word has a maximum of only one mis-keyed letter. The whole passage would be passed by a spell-checker, but it's utter nonsense:

Example of post-spellchecked garbage

His technique cam also by applies to the analyses or gold bills. The surface oh a gulf bell hat dimpled an is, ant whet is travels thorough aid the flop around the bell it smother.

Translated: This technique can also be applied to the analysis of golf balls. The surface of a golf ball has dimples on it, and when it travels through air the flow around the ball is smoother.

2 Have a list of commonly mis-spelt words to hand

Even if you think you can spell a word correctly, it pays to check.

Some examples of words that are commonly mis-spelt in science writing

Wrong	Right
accomodation	accommodation
alot	a lot
callibrated	calibrated
comparitive	comparative
consistant	consistent
equillibrium	equilibrium
guage	gauge
heirarchy	hierarchy
intergrate	integrate
proceedure	procedure
recomend/reccomend	recommend
rythm	rhythm
seperate	separate
speciman	specimen
theoritical	theoretical
verses	versus (as in describing a graph)
vise versa (and variations)	vice versa
yeild	yield

3 Use a dictionary

This is tedious if you know you're not a good speller.

Putting it right: spelling

If you know you're not a good speller, the best strategy is to try to word-process and spell-check your assignment. Then rigorously proof-read it.

Use small words, not big ones

By choosing a big word where a small one will do, we are trying to bluff people into thinking that we are impressive. Short words will make your writing fresh and vivid.

The list below contains pairs of words that mean the same thing. Technical and professional writers will almost invariably choose the word in the first column, and end up sounding pompous. Your writing will be more direct if you choose the shorter word. This is not to say avoid the longer words altogether. Avoid using them exclusively, and aim for a mixture of long and short. This will help your reader not to be bored.

Pompous word	Short word
anticipate	expect
assist	help
commence	start
desire	want
endeavour	try
indicate, reveal	show
locate	find
request	ask
require	need
terminate	end
utilise	use

Use appropriate technical terms

Use the technical terms appropriate to your discipline. But avoid pre-packed jargon. We have to distinguish between the jargon associated with our particular discipline, and pre-packed, clichéd jargon. The first is thoroughly necessary when writing about scientific or engineering concepts; one of the signs of an inexperienced report or essay writer is the lack of use of the appropriate terminology.

But pre-packed jargonese is another matter. Writing that contains phrases like *at this point in time* and *at the end of the day* are very irritating to read. Below is a list of some of the phrases to avoid.

a window of opportunity
all things being equal
as a last resort
as a matter of fact
at the end of the day
at this point in time
comparing apples with apples
conspicuous by its absence
easier said than done
effective and efficient
if and when
in the foreseeable future
in the long run
in the matter of
it stands to reason
last but not least
level playing field
many and diverse
needless to say
on the right track
par for the course
slowly but surely
the bottom line

Don't use *I* or *We* too often

In reports

It is best to limit the use of *I* and *We* in reports unless it is absolutely required, otherwise it can sound childlike. *We then opened Valve X* is easily turned into the passive *Valve X was then opened* (see page 135 for an explanation of the passive voice).

But sometimes, in trying to avoid using *I* or *We*, we can sound pompous. Occasional use of them will liven up writing. But you need to be aware of the dividing line between freshness of style and making it sound like a ripping good yarn (*What We Did in the Lab Today*).

In essays

Even though you may be required in an essay to develop an argument that is a statement of your beliefs, many lecturers will not accept *I feel* or *I believe* as valid ways of expressing yourself. They do not want an outpouring of your personal feelings; they are looking for a well-constructed argument that should appeal to the intellect.

Suggestions for avoiding *I feel* and *I believe*:

- *It may be concluded, therefore, that* ...or *We may conclude, therefore, that* ...
- *Clearly,* ...or *It is clear that* ...
- *We can reasonably suppose that* ...or *It can be reasonably supposed that* ...
- *The evidence supports the view that* ...or *The evidence suggests that* ...
- *It may be assumed that* ...or *We may assume that* ...

Use gender-neutral language

Language is considered sexist when it implies only one sex when both are intended. We therefore have to make sure that the language that we write and speak is as neutral in gender as possible. In contrast to only a few years ago, it is now seen to be inappropriate to use language that can be said to exclude women. By far the commonest cases are the use of **he/his** and **man/men** when both men and women are being referred to, as in this quote from The Environmental Pollution Panel of the President's Advisory Committee, 1965:

> These changes may affect man directly, or through his supplies of water and of agricultural and other biological products, his physical objects or possessions, or his opportunities for recreation or appreciation of nature.

Another area that usually promotes debate – often heated, with entrenched attitudes – is whether words such as manmade and mankind are non-sexist, in spite of the prefix. It is also necessary to remember that some words starting with *man-* such as *manual labour, manufacture* and *manipulate* are derived from *manus,* Latin for hand, and have nothing to do with maleness.

All students and staff – not only women – need to feel that they are in an educational system that is committed to its policy of promoting equity. No-one should feel excluded through a lack of thoughtful language. Careful writers avoid usages that give this sense of exclusion. It isn't necessary to make your writing clumsy by substituting *he or she* every time. Neither is there any need to go to the extremes of political correctness and embrace absurd uses such as *personhole cover*.

Ways of ensuring that your writing is more sensitive:

1 **Rephrase the sentence using inclusive or neutral words.** Instead of *man*, use *people, humans* or *human beings*.

Original incorrect version	Corrected version using neutral words
When man first discovered metals, he probably found by accident that a nodule of copper remained after his wood fire had gone out.	When humans first discovered metals, they probably found by accident that a nodule of copper remained after their wood fires had gone out.

2 **Rewrite using the plural:**

Original incorrect version	Corrected version using the plural
As far as the individual consumer is concerned, his electricity is cheap and therefore he will not want to install a solar heater. **Be sure that you haven't created a mixture of singular and plural, as in:** As far as the individual consumer is concerned, their electricity is cheap and therefore they will not want to install solar heaters.	As far as the individual consumers are concerned, their electricity is cheap and therefore they will not want to install solar heaters.

3 **Rewrite using the passive voice** (see page 135 for an explanation of the passive voice).

Original incorrect version	Corrected version using the passive voice
It was not until about 1500 BC that man perfected his iron working techniques.	It was not until about 1500 BC that iron-working techniques were perfected.

4 Use *he or she* or *she or he* sparingly
An occasional *he or she* (and the reverse order 'she or he') is acceptable, but a number of them in a short space destroys the sense of a passage. Avoid *he/she* or *s/he*. Both of these usages are clumsy.

Original incorrect version	**Corrected using he or she**
If a student is having difficulties with a course, should see his lecturer.	If a student is having difficulties with a course, he or she he should see the lecturer.

Write formally: avoid colloquialisms and contractions

1 **Slang and colloquialisms** should not be used in reports and essays.

2 **Contractions**
Contractions are where two words have been squashed into one: for example, *shouldn't, mustn't, wouldn't*. In an assignment, the words should be written out in full: *should not, must not, would not*. Contractions are informal and colloquial, so they should not be used in assignment writing.

Examples of contractions- don't use	**Example of how not to write**	**Rewrite it as:**
couldn't	The valve couldn't be opened.	could not
wouldn't	The young birds wouldn't feed.	would not
wasn't	The stream wasn't polluted.	was not
weren't	The older birds weren't present.	were not
didn't	The water didn't contain PCBs.	did not
shouldn't	This procedure shouldn't have been followed	should not
there's	There's no reason to suppose that ...	There is
it's	It's therefore reasonable to suppose that ...	It is
He's/he'll	He's said he'll consider the issue	He has said he will consider

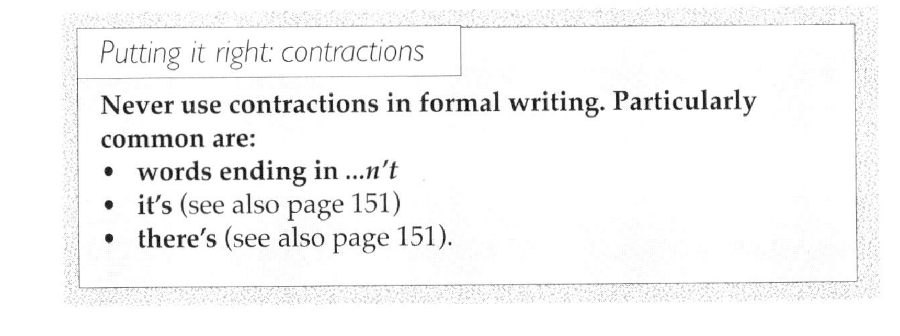
Putting it right: contractions

Never use contractions in formal writing. Particularly common are:
- **words ending in ...*n't***
- **it's** (see also page 151)
- **there's** (see also page 151).

See also page 151 **Using the apostrophe**

Choosing the right word: pairs of words that are often mixed up

There are pairs of words or expressions that are often muddled up. Choosing the wrong one can irritate your marker.

Some of the most common pairs are:
- absorb / adsorb
- affect / effect
- complement / compliment
- imminent / eminent
- it is composed of / it comprises
- its / it's
- lead / led
- loose / lose
- passed / past
- principal / principle
- their / there
- whose / who's.

The next section gives some easy ways of deciding on the right word.

1 Absorb/adsorb

Absorb means:
- to take up by chemical or physical action
- the swallowing up or engulfing of something.

Adsorb describes the process of the adhering of atoms or molecules to exposed surfaces, usually of a solid. It should by used only when you need this precise meaning.

2 Affect/effect

This is easy when you know how. Focussing on the commonest uses of the two words in undergraduate science writing:
Affect is a verb, **effect** is a noun (see Appendix 1 if you need help with the terms).

To **affect** something is to influence it (*a verb*):

- The pollution will affect the dissolved oxygen concentration.
- The pollution has affected the dissolved oxygen concentration.

The **effect** of something is the result or consequence of it (*a noun*):

- The pollution will have a great effect on the dissolved oxygen concentration.

Incorrect	**Corrected**
There are natural events, such as volcanoes or eruptions, which may *effect* the ozone layer.	There are natural events, such as volcanoes or eruptions, which may *affect* the ozone layer.
Most of the countries not *effected* by acid rain are doing little to decrease emissions.	Most of the countries not *affected* by acid rain are doing little to decrease emissions.
In some systems the fish were unable to be eaten, therefore *effecting* the food chain.	In some systems the fish were unable to be eaten, therefore *affecting* the food chain.
This report examines the *affects* of natural hazards on communities.	This report examines the *effects* of natural hazards on communities.

Putting it right: affect/effect

Effect: Use *effect* or *effects* as nouns. This means:
- *a, an* or *the effect*
- *the, some*, or words meaning more than one (*several, a few, many, a couple of*, etc.) *effects*.

Never use: this will effect the stream, this has effected the stream

Affect: You can only use *affects* as a verb.
This *affects something*, this *will affect something*, this *has affected something*.

Never use: the *affects* of this

Affected will almost certainly be what you need, ***not*** **effected**.

3 Compliment *and* complimentary/Complement *and* complementary

Compliments/complimentary

Where you want to imply flattery:

- She complimented him on his presentation.
- We give you a complimentary bottle of champagne.

Complements/complementary

a The finishing touches to a thing, fitting together, completing;
The formal garden complements the exterior of this superb house.

b The scientific or mathematical meanings:

- complementary angles
- complementary colour
- complementary relationship
- complementary function

4 Imminent/eminent

Imminent: soon, impending:
Their arrival is imminent

Eminent: important, distinguished:
She is an eminent scientist

Incorrect	Corrected
The extinction of this species of bird is *eminent*.	The extinction of this species of bird is *imminent*.

5 It is composed of /it comprises

There are two expressions that get mixed up: **is composed of** and **comprises**.

You can say 'It is composed of three parts' or 'It comprises three parts'. But you can't say 'It is comprised of three parts'. This is a common mistake.

Incorrect	Corrected
A diesel engine is like any other modern engine in that it is *comprised of* a cylinder block and a crankshaft.	A diesel engine is like any other modern engine in that it *comprises* a cylinder block and a crankshaft. *or* A diesel engine is like any other modern engine in that it is *composed* of a cylinder block and a crankshaft.

Putting it right: is composed of/comprises

Comprised or *comprises* can never be followed by *of*.

6 Lead/led

This has become confused because *lead* is pronounced in two different ways:

- the element lead (Pb)
- Will you lead the team?

The most common mis-use is in:

Incorrect	**Corrected**
This *lead* to pollution of the stream.	This *led* to pollution of the stream.
This has *lead* to more interest being shown in the hot air engine.	This has *led* to more interest being shown in the hot air engine.

Putting it right: lead/led

Whenever you write *lead* say it to yourself.

- Does it sound like *led*? If so, write **led** (unless you mean the element lead, Pb).
- Does it sound like *leed*? If so, write **lead**.

This word doesn't follow the same system as *read*, which is what often confuses people.

7 Lose/Loose

In their most usual senses in science writing:
Lose means to cease to possess or misplace.

Loose means not restrained:
A **loose** fit. The cover was **loose**.

Incorrect	**Corrected**
The breeding pairs will *loose* their chicks if conditions do not improve.	The breeding pairs will *lose* their chicks if conditions do not improve.

8 Passed/past

Passed: often means accepted or carried
- The law that has just been passed states that ...
- Somatic injury is not passed on to the next generation

Past: can mean just gone by and refers to time
- In the past
- Past practices
- Over the past year

9 Principal/principle

Principal: the most important, the highest in rank, the foremost
- The study was made up of five principal sections.
- The principal of the institution said that...

Principle: a fundamental basis of something
- Archimedes' principle.
- The minister has no principles.
- The principles of the investigation were...

10 Their/there

- If you ever find yourself writing:
  Their was, Their is, Their could be/should be/would be, Their will write **there** instead. **Their** is never followed by *is, was, will, should, would*, or *could*.

- On all other occasions (except when you are saying something is *over there*) you are likely to need *their*.

Incorrect	**Corrected**
Their are a number of strategies that countries can take.	*There* are a number of strategies that countries can take (You can't have *their* and *are* together).
The Phillips Company was trying to find a power source for *there* radios.	The Phillips Company was trying to find a power source for *their* radios.
In *there* advanced form they are superior to petrol engines.	In *their* advanced form they are superior to petrol engines.
Small boats will break from *there* moorings.	Small boats will break from *their* moorings.

Don't write breathlessly

Excited writing and exclamation marks are not appropriate in balanced writing. Avoid phrasing such as *The world has a problem with carbon dioxide, of that we can be sure! There is now the possibility of restoring these sites back to their original (hopefully!) condition.*

Some other aspects of style

Using the apostrophe

There are two areas where an apostrophe is used:
- **the possessive** – denoting who or what something belongs to
- **contractions** – where two words have been informally squashed into one.

1 The possessive

Here the apostrophe shows who owns something.

Use **'s** if there is only one:	the cell's chromosomes (the chromosomes in one cell)
Use **s'** if there are more than one:	The cells' chromosomes (the chromosomes in all the cells)

Note, though, that there is no apostrophe in *yours* (e.g. the book is yours), *hers, ours, theirs* or in *its* (e.g. the cell and its chromosomes, *see also next section*).

2 Contractions

The apostrophe is used to show that two words have been informally pushed together. Since they are informal, they should never be used in your assignments. They are a source of great confusion, but are easy when you know how.

The main contractions are:
- it's
- who's
- there's
- everything ending in *..n't* (wouldn't, hadn't etc.).

See also page 145 – the section **Writing formally: colloquialism and contractions.**

It's/its

Never use it's in any assignment

The mixing up of these two happens all the time, yet it is very easy to understand the difference.

It's is the short, colloquial way to write *it is* or, less often, *it has*. You don't use colloquial forms in writing reports and essays.

Incorrect	**Corrected**
Unless large amounts of fertiliser are spread the land loses all it's nutrients.	Unless large amounts of fertiliser are spread the land loses all its nutrients.

Putting it right: it's/its

- Never write *it's* in a report or essay.

Whenever you are writing something formal (i.e. something other than informal letters or notes), you will need the *its* form – with no exceptions.

A good way to check: read it aloud to yourself, reading every *it's* as *it is*. Does *it is* make sense? If not, write *its*.

Whose/who's

This case is very like *it's/its*. You never write *who's* in assignment writing. This is because it's a colloquial form of *who is*.

Incorrect	**Corrected**
Mr. Smith, who's responsibility is the monitoring of the outfall, says that...	Mr. Smith, whose responsibility is the monitoring of the outfall, says that... (Does it mean *who* is? No: so write *whose*.)
Mr. Smith, who's responsible for monitoring the outfall, says that...	Mr. Smith, who is responsible for monitoring of the outfall, says that... (Does it mean *who is*? Yes: so write *who is*.)

Putting it right: whose/who's

- If you mean *who is*, write it.
- All other times, you'll need *whose*.

Plurals

Plurals are **never** made by using an apostrophe.

Putting it right: plurals

- Never make a plural – more than one of something – by adding *'s*. Just add *s*. (Forget about tomatoes and potatoes).

- Therefore you can never have:
  Some river's, some plant's, some valve's.
  It has to be some rivers, some plants, some valves.
- Not even the plurals of abbreviations have apostrophes.
  Correct: CFCs, PCBs, CDs.

Irregular plurals

Words commonly used in science and which have irregular plural forms are:

Singular	Plural
alga	algae
analysis	analyses
antenna	antennae (zoology) antennas (communications engineering)
appendix	appendices
axis	axes
bacterium	bacteria
criterion	criteria
datum	data
formula	formulae
genus	genera
hypothesis	hypotheses
larva	larvae
matrix	matrices
medium	media
nucleus	nuclei
ovum	ova
phenomenon	phenomena
quantum	quanta
radius	radii
species	species
stimulus	stimuli
stratum	strata
symposium	symposia
vertebra	vertebrae
vortex	vortices

The split infinitive

Split infinitives aren't nearly as important as many people make them out to be. There isn't, as many people suppose, a rule that says an infinitive should not be split; it is merely an invention observed by them in the mistaken belief that they are showing their knowledge of 'good' writing. Rigorously sticking to such outmoded ideas does just the opposite; it interferes with your ability to communicate effectively. Only misinformed pedants criticise a piece of writing because it contains a split infinitive.

What is an infinitive? This is a verb form. When *to* is followed by a verb (the 'doing' word), it is said to be an infinitive: *to walk, to run, to think* etc.

A split infinitive is when words come between *to* and the verb:

They were urged to *seriously* reconsider their stand.

Pedants would insist that this sentence be rewritten. However, the three rewrites sound awkward:

They were urged to reconsider seriously their stand.
They were urged seriously to reconsider their stand (*this is also ambiguous*)
They were urged to reconsider their stand seriously.

However, a lengthy interruption can justifiably be criticised:

The political will is lacking to *resolutely, wholeheartedly and confidently* reform the tax system.

Sometimes, avoiding a split infinitive results in ambiguity:

He would like to *really* learn the language.

The alternatives are ambiguous: *He would like really to learn the language* could mean the same as *He would really like to learn the language*.

Putting it right: split infinitives

Write whatever sounds the least awkward. Only misinformed people worry about split infinitives.

However, you need to be aware that some markers fall into this category. They will hunt them down and delight in pointing out each one.

Numbers

These are the conventions for dealing with numbers in technical writing:

1 All important measured quantities should be expressed in figures. In particular this includes decimal points, dimensions, degrees, distances, weights and measures.

For example: 2.4 seconds, $5000, 5 °C, 9 kilometres, 6 tonnes, 3 amps

2 Counted numbers of ten or less are written out. Those above ten are numeralised:

Measurements were taken in five areas.
Measurements were taken in 20 areas.

3 A number at the beginning of a sentence should be written out:

Twenty samples were taken.

Commas, semicolons and colons: the mechanics

Note: The use of **quotation marks** in referencing is covered in Chapter 13 **References**, page 114.

Here are very brief guidelines on the main ways in which commas, semicolons and colons are used:

When do you use a comma?

A comma indicates a pause. You can often tell where a comma should be by saying the words to yourself. The places where commas are generally used are:

1 **After each item in a series, but generally not before the final *and*.**

Adjectives	The river is wide, turbulent and muddy.
Nouns	The most common birds on the island are sparrows, chaffinches, thrushes and blackbirds.
Phrases	The river-mouth is wide, with large shingle banks, extensive sand dunes and a small island.

2 **To delimit a subclause from the main clause in a sentence** (see page 128 for an explanation of clause and subclause).

Increasing agriculture will cause an increase in global warming, the reason being that ruminants and paddy fields produce methane.

When the engine was run on petrol the carbon dioxide emissions were higher, which was an indication of improved mixing.

3 **After an introductory phrase or subclause.**

Although farmers have reduced their use of pesticides in this area in recent years, there is still local concern about the issue.

By using better management practices, farmers have been able to reduce their use of pesticides.

4 **To delimit material that is not essential to the meaning of the sentence.**

The island, although windswept, has a large number of different bird species.

When do you use a semi-colon?

1 **Between two closely-related independent clauses.**
The statement on each side of a semicolon should be able to stand alone as a sentence (see also page 131 **Comma splices**).

In December 1952 a deadly smog hung over London for almost a week; it killed about 4000 people.

2 **Between groups of items in a list when the items are punctuated by commas.**

Yesterday I ate muesli, bacon and eggs for breakfast; bread, cheese and pickles for lunch; and fish and chips for dinner.

When do you use a colon?

To introduce a list or series.
The following topics will be discussed:
- global warming
- ozone depletion
- volcanic hazards.

The following topics will be discussed: global warming, ozone depletion and volcanic hazards.

Formatting equations in the text

There are minor variations in styles of formatting equations. The following shows a good general style:

The value of the shear stress at a distance r from the axis is given by

$$\tau = Gr\,\frac{d\phi}{dx} \qquad (3.5)$$

We can see from Eq. (3.5) that the shear stress acting on the circular cross section is linear in the radius r.

Notes

Equation is centred

Equation number in brackets is tabbed to the right margin. This is equation number 5 in Section 3 of the report.

In the text refer to the equation as either 'Eq. (*equation number*)' or 'equation (*equation number*)'. Be consistent in your use of one or the other throughout your text.

For a sequence of equations in which the left-hand side is unchanged:
Align the = symbol in each line.

$$\begin{aligned} u(x) &= -\frac{q_0}{AE}\int_0^x (x-\xi)\,d\xi + \frac{C_1 x}{AE} \\ &= -\frac{q_0 x^2}{2AE} + \frac{C_1 x}{AE} \end{aligned}$$

For continued expressions in which the left side is long:
Align the = symbol with the first operator in the first line.

$$\begin{aligned} &[(a_1+ia_2)+(a_{11}s_1+a_{21}s_2)]/[(b_1+ib_2)+(b_{11}s_1+b_{21}s_2)] \\ &\quad =f(x)g(y)+\ldots \end{aligned}$$

For expressions in which the right-hand side is long:
Align the continuing operator with the first term to the right of the = symbol.

$$\begin{aligned} V(x) = &-P\langle x\rangle^0+P(x-a)^0 \\ &+P\langle x-(L-a)\rangle^0-P\langle x-L\rangle^0+C_1 \end{aligned}$$

Genus and species names

The name of a species is in two parts, consisting of two Latin names: a genus name and a species epithet, e.g. *Rosa acicularis*. The conventions for the more simple aspects are given here. Use a style manual (see Further Reading) for greater detail.

- The initial letter of the genus name is capitalised.
- The initial letter of the species epithet is always in lower case (*Note:* it is a common mistake to capitalise it).
- The whole name is italicised (or underlined if you are writing by hand).
- A genus name should always be followed by a species epithet or, if the species is unknown, by 'species', 'sp' (singular) or 'spp' (plural), none of which are italicised, e.g. *Rosa* sp.
- A genus name should be spelled out on first mention in the text. Thereafter it can be abbreviated to the initial letter followed by a full-stop and the species epithet, e.g. *R. acicularis*.
- A variety is written as follows: *Rosa acicularis* var. *rotunda*.

Checklist: Problems of style and how to correct common mistakes

Write good sentences

- Disentangle long sentences.
- Avoid sentence fragments. Check if at the beginning of the sentence:
  - ✓ *...ing* has been used: two methods of correcting it.
  - ✓ *Which* has been used (when it's not a question): two methods of correcting it.
- Avoid comma splices: three methods.
- Use linking words to join bitty sentences.

Use verbs well

- Use active or passive voice according to the emphasis of the sentence.
- Avoid the distorted passive.
- Use the tense appropriately and consistently.
- Make subject and verb agree.

Get the words right

- Use the correct spelling:
  - ✓ use the spell-checker.
  - ✓ use a list of common words.
- Use small words, not big ones.
- Avoid pre-packed jargon. But use the terminology appropriate to your discipline.
- Use gender-neutral language.
- Don't use contractions or colloquialisms.
- Use the right word or phrase of a pair (*affect/effect, comprises/is composed of, lead/led etc.*).

Using the apostrophe

- Never write *it's* in an assignment.
- Never make a plural with an apostrophe.

- **Irregular plurals:** have a short list to hand

- **Be aware that pedants will hunt down each simple split infinitive,** even though there is nothing wrong with them.

- Use the accepted conventions for:
  - ✓ commas, semicolons and colons
  - ✓ equations
  - ✓ genus and species names.

Writing to get a job

Writing a CV and job letters

16 CHAPTER

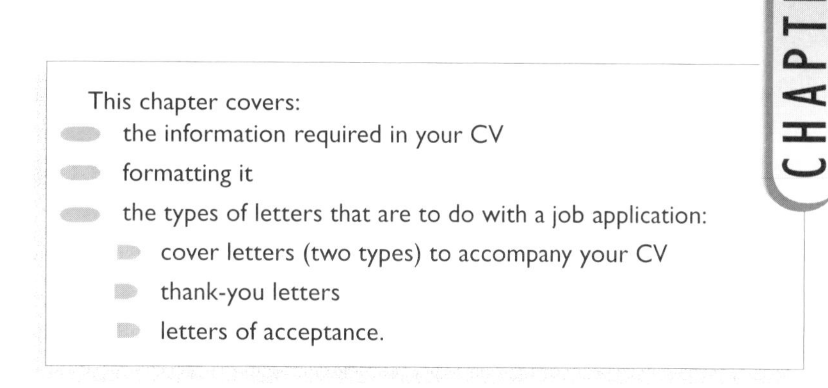

This chapter covers:

- the information required in your CV
- formatting it
- the types of letters that are to do with a job application:
  - cover letters (two types) to accompany your CV
  - thank-you letters
  - letters of acceptance.

Writing a CV

Writing a curriculum vitae (CV) is an area where you should follow the accepted models as closely as you can. You can add minor variations to show your individuality, but do stick to the proven conventions. Your readers in this case may expect the conventional format and you can confuse and irritate them if you deviate from it.

How long should it be?

Your CV has to make an impact in a much shorter time than you would think; management consultants who screen people for the first interview say that the average time spent on scanning a CV is 20 seconds.

Your CV at this stage of your career should be no more than 2–3 pages long. This is important. The worst thing you can do is to bulk it out with unnecessary information and poor formatting.

Here are the guidelines for writing a CV:

Formatting

- You are aiming for visual balance and easy scannability.
- No longer than two to three pages.
- Typed text is essential; good quality laser or ink-jet printing is preferable.
- Keep at least 2.5 cm margins on all four sides of the page.
- Indent the text so that you have variable margins. This makes for easier scanning. But keep identical margins for related

information (i.e., line up your major sections with each other; line up your job specifications etc.).

- Don't format it so that you have enormous areas of white space with just a few items of information on a page. Aim to present your information in a dense, but pleasingly formatted way.
- Don't use too many different fonts. You can enhance your CV by using one font (say a sans-serif font such as Arial or Helvetica) for headings and another (say a serif font such as Times Roman) for text. Any more than that and it looks a mess.
- You can use different point sizes and boldface text to effect, but only if it is done with discretion and sense. It is all too easy to end up with a mess.
- Be consistent in your use of fonts, point sizes and other formatting. If you boldface one job title, boldface them all.
- Have someone else proofread it for you. It has to be immaculate – many applications are passed over because of grammatical and spelling errors in the CV.
- Present it on white paper. Avoid coloured or hand-made paper.

Content

Sections to include in a CV
In order:

- Personal details
- Education
- Work experience
- Specialised skills
- Interests
- Personal attributes (if wanted)
- Career Goal (if wanted)
- Referees

Personal details

Name, address, telephone number. Your birth date, marital status and state of health are optional.

Education

- Start with your most recent year of study and work backwards. **This is important. Don't start far back and work towards the present.**
- For university papers, include the title, not just the paper number (this by itself won't mean anything to an employer).
- Putting grades is optional, but most employers will want to see them at some point. You don't want to look as though you are hiding something, even if your grades are not as good as you want.
- If your academic record is patchy or broken, you may need to include a positive statement of explanation.
- Include your final school year results.

- You can include earlier school results, particularly if they show extension, acceleration, or early signs of academic achievement.
- Don't forget to include other tertiary qualifications.

Work experience

- Start with your most recent job and work backwards.
- Include all your part-time jobs, even if they were very short term. If you have had a large number of part-time jobs you could group them together. Make sure you include all your jobs, even if they seem trivial; employers look for evidence of adaptability.
- Include a very brief description of your duties and responsibilities, except for part-time jobs of very short duration.
- Include evidence of any initiatives taken, extra responsibility offered, and supervision of others.

Interests/Other activities

Subheadings you could include are: Cultural, Sporting, Social and Community, Skills and Experience. Remember that employers are looking for a sense of you as a person – in particular, evidence of achievement and personal skills – no matter if you're a back-room boffin or a proactive type. So don't just put 'music' if you are a skilled French horn player, or 'diving' if you have an advanced PADI qualification. Give some brief details. If reading is a main interest, say what types of books you like.

Specialised skills

Include skills such as competence with specific types of software, practical skills such as machine-shop skills, etc.

Personal attributes/Career goal

Much advice on CV-writing suggests that you include a statement about your personal qualities and where you see yourself in five years' time. Later in your career you will have the background to be able to do it effectively; at this stage it is not a good idea. Your employers will be looking for information that stands up to scrutiny; statements such as your enjoyment of challenge, or that you are well-organised only come across as being murky.

However, if you can provide supporting evidence and can elaborate it at interview with specific examples, you could mention, for instance, that you work well in a team and support it with work and/or community, cultural group or sporting experience.

References

- You need to be able to nominate two or three people as referees. Ask people who know you reasonably well. Try to include someone who is familiar with your personal qualities; someone – preferably a person connected with a vacation job – who can speak about your work skills; and, if possible, a university member of staff who knows your academic capabilities. Many students and recent graduates are not in a position to have this last type of referee. Don't worry about it – it doesn't matter.

- Include telephone numbers for each of your referees. Employers often prefer to phone rather than write.
- Ask these people if they will act as referees for you, give them a copy of your CV, and remember to tell them when you have applied for a job.

Example of a CV for someone about to graduate:

Sarah Elizabeth Andrews
Curriculum Vitae

Personal details

Name	Sarah Elizabeth Andrews
Address	154 King Street Eastleigh Oakfield 6
Telephone	(09) 412 7834 Secondary number for messages: (09) 465-7784
Date of birth	7 September 1975
Citizenship	British
Health	Excellent, non-smoker

Education

1994–1996, The University of Middletown

1996	**Bachelor of Science (Biological Sciences)**	
	Conservation Ecology and Genetics	
	Freshwater and Estuarine Ecology	
	Biological Oceanography	
	Ecological Physiology	
	Coastal Marine Ecology and Aquaculture	
	Pure and Applied Plant Development Biology	
1995	Animal Function and Design	B
	Genetics	B+
	Principles of Ecology	C+
	Biochemistry	B-
	The Biology of Marine Organisms	A
	Biometry	A-
	Cellular and Molecular Biology	B-
1994	Central Concepts in Biology	A-
	Principles of Computing (a)	A-
	Principles of Computing (b)	B+
	Principles of Statistics	A-
	Introduction to General Psychology	C+
	Organic and Physical Chemistry	B+

1989–1993, Eastleigh High School

1993	**University Entrance**	
	Biology	A
	Physics	B
	Chemistry	C
	English	B
	Mathematics	A
1992	**Sixth Form Certificate** English, Mathematics, Physics, Chemistry, Geography	
1991	**School Certificate** English, Mathematics, Science, Geography, History	
School activities	Student representative on School Council Form representative Represented school in badminton (7th form) and tennis (6th form) Duke of Edinburgh award (silver)	

Work Experience

Nov 1995–Feb 1996	Field assistant, Department of Conservation. Responsible for monitoring Chaffinch numbers on West Island, and compiling database records.
Nov 1994–Feb 1995	Storesperson, Machinery Hire Ltd, Westport. Responsible for inventory control and filing.
May 3–25 1994, Aug 14–Sept 7 1994	Counter assistant, Liz's Hamburgers Ltd, Eastleigh.
Nov 1993–Feb 1994	Tomato and raspberry picking, McDonald Growers Ltd, Eastleigh.

Specialised skills

Experienced with the software packages SAS, Microsoft Word, Microsoft Excel.
Diving qualification: PADI open-water diving certificate.

Interests

Diving (PADI-qualified open water), aerobics, Royal Forest and Bird Society area committee member, reading (modern history, modern novels).

Referees

Mr. Peter McCarthey JP, 45 North Street, Eastleigh. Phone: (09) 435-9886
Mr. Jim Wilson, Machinery Hire Ltd, Westport. Phone: (09) 644-7622
Dr. Margaret Horley, School of Biological Sciences, The University of Middletown.
Phone: (06) 334-9264

Writing letters of application

When you are looking for a job, the types of letters that you may have to write are:

1. A **cover letter for a permanent position**. This introduces your CV and asks for an interview. It can be of two types:
   - In response to an advertisement.
   - Not in response to an advertisement, but writing to a company enquiring if there is a possibility of employment.
2. A **cover letter** enquiring about **vacation employment**.
3. A **thank-you letter**, expressing thanks for an interview or an invited visit.
4. A **letter of acceptance**, accepting a job offer.

Here are the overall guidelines for writing letters:
- Limit letters to one page if possible.
- Use single-spaced or 1.5 spacing.
- Leave 2.5 cm margins on all four sides of the page, and set the complete letter on the page so that it is visually balanced, not crammed in the top half of the page.
- Leave one blank line between paragraphs.
- Do not use fancy fonts. Be completely conventional.
- Proofread very carefully.

The heading and greeting

It is now standard practice to left-align the two addresses, the date and the greeting. Put your address and telephone number at the top of the page. Under it put the date. Then under that put the name and address of the person to whom you are writing.

> **Important note:** Always address a letter to a particular person by name in the company, not just to a position such as The Human Resources Officer. If you don't know their name, ring up the receptionist and find it out, together with their title and gender. This shows to the company that you have taken the initiative to do this. It is far more impressive than an impersonally addressed letter.

Then skip a line or two and write *Dear*, the person's title (*Mr., Ms, Dr.*), their name, and a comma.

The ending

- Keep your closing short and simple, gracious and sincere. Don't be too pushy or falsely flattering. If it is relevant, don't forget to indicate your contact phone numbers if you are likely to be rung up over a break or during the summer.
- Under the final paragraph, skip a line or two and write 'Yours

sincerely,'. Skip 6–8 lines then type your name. In the space left by the skipped lines, sign your name.

Note on convention: The standard conventions – not always observed – for ending a letter are:

- if your letter started Dear Sir or Dear Madam, end it with 'Yours faithfully'.
- if it started by using the person's name – Dear Mr Smith, Dear Ms Smith, Dear Dr. Smith – end it with 'Yours sincerely'.

- If you are including something with your letter (such as your CV), write the word 'Enclosure' at the bottom of the page.

The Cover Letter Replying to an Advertisement

Notes:

In the first paragraph state the specific position that you are applying for, with the job title and the vacancy number if there is one. State where the position was advertised.

In the second paragraph state:

- Why you are interested in the position and in the specific organisation.
- What you have to offer the organisation (it pays to have done some investigation into what the organisation does).
- The relevance to the advertised position of your academic record, work experience and abilities. Keep your skills and strengths clear in your mind.

In the third paragraph refer to your enclosed CV. You do not need to refer to your academic record if you do not want to.

In the final paragraph ask for an interview and show your willingness to expand on the information contained in the CV and letter.

Sample

413 Mount Victoria Rd
Epsom
Middletown

12 October 1995

Mr L. J. Carter
Peterson Associates Ltd.
125 Great West Road
Middletown

Dear Mr. Carter,

I would like to apply for the position of Environmental Planner within the Environmental Planning Section of Peterson Associates Ltd (Vacancy Number AF/34), which was advertised in the *Independent* on Tuesday 14 March 1996.

I am very interested in developing my career as an environmental planner, in particular with an interdisciplinary team. My degree includes many papers that are directly relevant to the position advertised, and during my undergraduate years I have been actively involved with environmental issues within small communities. In particular I have an excellent understanding of the legislative procedures. I completed two research projects that required a detailed understanding of the legal aspects of the issues. One of these projects, described in my attached CV, was for the Burleigh Community Board, which took place over three months. It resulted in a detailed written report and a presentation to three community groups.

As you can see from my CV I have a good academic record, wide-ranging interests and enjoy working with people. I am keen to continue my involvement and study in this area.

I would very much like an opportunity to discuss my application more fully with you. I am available for an interview at any time that is convenient for you. My telephone number is (039) 453-6824; messages can also be left for me on (039) 584-8834.

I look forward to hearing from you.

Yours sincerely,

Rebecca Fowles

Enclosure

The blind cover letter accompanying a CV, enquiring if the company has a possible opening for you

In the first paragraph state what degree you are doing, any specialisations within it and your particular areas of interest.

In the second paragraph say why you are interested in this particular company.

In the third paragraph state how you think you can be of particular use to the company in terms of your academic record, work experience and abilities. These linkages are the key to selling yourself to the company.

In the final paragraph state that you are enclosing your CV, and that you welcome an interview and the opportunity to expand on the information you have given. Make sure that you can be readily reached on the telephone numbers that you give.

Sample

Dear Ms. King,

I am in my final year of an Honours degree in Biological Sciences, specialising in Coastal Marine Ecology and Aquaculture. I have a particular interest in the monitoring of pollution, and am completing a research project on the possible effects of tri-butyl tin on a population of cultivated mussels.

I am writing to enquire whether there is a possibility of employment with BioResearch Methods Ltd. In the course of my studies I have on a number of occasions referred to some of your research reports, and I am convinced that your organisation's work is exactly of the type that I would like to pursue. It has the strong analytical aspect that I most enjoy, and I would very much like to be involved in your input into major environmental issues.

My degree has given me a strong background in the practical aspects of marine ecology. As part of my Bachelor of Science degree, I have taken two special courses in the management of shallow water marine systems and a course in the culture of benthic marine organisms. I have also had summer experience as field assistant to Professor Roger Woods of the Bayview Marine Research Centre, where I was monitoring experimental mussel beds for a three-month period. I am a PADI-qualified advanced scuba diver, and have also been actively involved with environmental organisations and community groups for five years. I believe that I can enrich my analytical skills with BioResearch Methods.

I enclose my CV. I am readily available to meet you and discuss possible opportunities, and I would welcome an interview. If you need further information, I can be reached on (067) 564-8122. If you are unable to contact me at that number, my parents can be reached on (05) 499-3346. I look forward to hearing from you soon.

Yours sincerely,

Mark Jackson

Enclosure

The blind letter enquiring for a holiday job

Dear Ms. Long,

I am writing to enquire about possible vacation employment for the months of December 1995–February 1996.

I am completing my first year in Mechanical Engineering in The University of Middletown. As my enclosed CV shows, I have a thorough background in mathematics, chemistry and physics, together with introductory design and computing. I am enthusiastic about applying this background to vacation work in a manufacturing company. My degree has a requirement for 800 hours of engineering-related work, and I am particularly keen to work for a company such as Plastic Extrusions Ltd, since my primary interest is in plastics manufacturing processes.

I would very much like to discuss my background with you at a time convenient to you. Dr. Paul Morris (09 425-7449 Ext 5874), senior lecturer in Mechanical Engineering, is also happy to speak to you about me. My daytime phone number is 09 745-4473, and a secondary number where you could leave a message is 09 454-5683.

Thank you for your consideration.

Yours sincerely,

Louise Robinson

Enclosure

A Thank-You Letter, expressing thanks for an interview or an invited visit

Dear Mr. Sharp,

Thank you for taking the time to speak to me on Monday, April 3. It was obvious from all the activity that you had a full schedule, and I appreciate your fitting me in so readily.

As our discussion made clear, my primary interest is in microbiological remediation. I share your belief that a microbiologist is needed in your team. My work over the past few years has convinced me that microbiology will play a primary role in the future remediation of pollution, and the prospect of working in an interdisciplinary consulting team is particularly exciting.

I look forward to hearing soon from you about Barrett International's hiring plans, and I would like to thank you again for all your time and attention.

Yours sincerely,

Nicholas Brown

A letter of acceptance of a job offer

Dear Ms. Allen,

It is a pleasure to acknowledge your letter of August 5 offering me a position as a process engineer with the Gelato Icecream Company. I am delighted to accept.

I understand that the conditions of employment require that I complete my Bachelor of Engineering degree in October and pass a medical examination.

As you suggested, February 1 1996 is a convenient starting date for me, and I am very happy to accept your offer of two weeks temporary housing. Gelato has, as always, been extremely hospitable in its dealings with me, and I look forward very much indeed to a challenging and rewarding career with you.

Yours sincerely,

Joanna Wesley

Checklist: Writing a CV and job letters

CV

- ✓ 2–3 pages long. Don't bulk it out.
- ✓ Format it for readability. But don't spread the information too thinly on the page.
- ✓ Proof-read it meticulously.
- ✓ Content: the usual sections are: **Personal details, Education, Work experience, Specialised skills, Interests, Personal attributes (if wanted), Referees.**
- ✓ Start with the present time and work backwards.

Job letters

There are four classes:

1. A **cover letter** for a permanent position. This introduces your CV and asks for an interview. It can be of two types:
   - ✓ In response to an advertisement.
   - ✓ Not in response to an advertisement, but writing to a company enquiring if there is a possibility of employment.
2. A **cover letter** enquiring about vacation employment.
3. A **thank-you letter**, expressing thanks for an interview or an invited visit.
4. A **letter of acceptance**, accepting a job offer.

APPENDICES

Appendix 1: the parts of speech

Parts of speech	The work that words do in a sentence
verbs	Words that indicate action: what is done, or what was done, or what is said to be. The ship *sailed*
nouns	Names. Things. *Columbus* sailed in the *ship*
pronouns	Words used instead of nouns so that nouns need not be repeated. *He* sailed in *it.*
adjectives	Words that describe or qualify nouns. The *tall* man sailed in the *big* ship.
adverbs	Words that modify verbs, adjectives and other adverbs. They end in *-ly*. The big ship *slowly* sailed past the *steeply* sloping cliffs.
prepositions	Each preposition governs, and marks the relation between, a noun or pronoun and some other word in the sentence. The ship sailed *past* the cliffs and *across* the sea *to* America.
conjunctions	Words used to join the parts of a sentence, or make two sentences into one: *and, but, so, because, as, while, since.* • The ship sailed to America *and* came straight back. • The ship sailed to America *but* did not stay long. • The ship sailed fast, *so* it got there quickly. • The ship sailed slowly *because* the sails were torn.
gerund	A word ending in *-ing* that behaves in some ways like a noun and in some ways like a verb. She likes *using* a computer You can save electricity by *switching* off the lights.

Appendix 2: tenses of the verb

This section describes, in very simple terms, the main ways of expressing tense in a verb.

Present tense

Describes what is happening at the moment:

1 The sun *shines*.
  Chlorofluorocarbons *cause* ozone depletion.
2 The sun *is shining*.
  Acid rain *is developing* into a major environmental issue

Past tense

Describes what happened in the past.

1 The sun *shone*.
  The eruption *caused* huge losses of crops.
2 The sun *was shining*.
  It was found that the amount of waste *was causing* blockage of the system.
3 The sun *has shone*.
  The use of chlorofluorocarbons *has caused* ozone depletion.
4 The sun *had shone*.
  By the 1950s, carbon dioxide levels in the atmosphere *had risen* to 315 ppm.

Future tense

Describes what is going to happen in the future.

1 The sun *will shine*.
  Increased emission of greenhouse gases *will cause* a change in the global climate.
2 The sun *will be shining*.
  By the middle of next century, the increased emission of greenhouse gases *will be causing* a global change in climate.
3 The sun *will have shone*.
  By the middle of next century, carbon dioxide levels *will have risen* to twice the pre-industrial level.

Appendix 3: units and their abbreviations

(Adapted from *Scientific Style and Format: the CBE Manual for Authors, Editors and Publishers*, (1994)

SI base units and symbols

Quantity	Name	Symbol
Base units		
length	metre	m
mass	kilogram	kg
Commonly used unit of mass:	gram	g (not gm)
time	second	s
electric current	ampere	A
temperature	degrees Kelvin	°K
	degrees Celsius (acceptable for experimental temperatures)	°C
volume	cubic metre	m^3
Commonly used unit of volume:	cubic centimetre	cm^3 (not cc)
amount of substance	mole	mol
luminous intensity	candela	cd
Supplementary units		
plane angle	radian	rad
solid angle	steradian	sr

Other units used with SI

Name	In terms of other units	Symbol
atmosphere	101 325 Pa	atm
calorie	4.18 J	cal
day	24 h	d
degree	(π/180) rad	°
hour	60 min	h
kilogram-force	9.8067 N	kgf
litre	1 dm^3	l
micron	10^{-6} m	μ
minute	60 s	min
minute	(π/10 800) rad	'
angular second	(π/648 000) rad	"
tonne	10^3 kg	t
torr	133.322 Pa	torr

Examples of SI derived units

Quantity	Name	Symbol	In terms of other units
Activity of a radionuclide	becquerel	Bq	s^{-1}
Acceleration			m/s^2
Capacitance	farad	F	C/V
Current density			A/m^2
Electric charge, quantity of electricity	coulomb	C	s A
Electric potential, electromotive force, potential difference	volt	V	W/A
Energy, work, quantity of heat	joule	J	N m
Energy density			J/m^3
Force	newton	N	$(m\ kg)/s^2$
Frequency	hertz	Hz	s^{-2}
Heat capacity, entropy			J/K
Illuminance	lux	lx	lm/m^2
Luminance			cd/m^2
Luminous flux	lumen	lm	cd sr
Magnetic flux	weber	Wb	V s
Moment of force			N m
Power, radiant flux	watt	W	J/s
Pressure, stress	pascal	Pa	N/m^2

Standard prefixes used with SI units

A prefix is a verbal element used before a word to qualify its meaning, eg *milli*metre (mm) – a thousandth of a metre; *kilo*metre (km) – a thousand metres; *milli*litre (ml) – a thousandth of a litre, etc.

Term	Multiple	Prefix	Symbol
10^{24}	1 000 000 000 000 000 000 000 000	yotta	Y
10^{21}	1 000 000 000 000 000 000 000	zetta	Z
10^{18}	1 000 000 000 000 000 000	exa	E
10^{15}	1 000 000 000 000 000	peta	P
10^{12}	1 000 000 000 000	tera	T
10^{9}	1 000 000 000	giga	G
10^{6}	1 000 000	mega	M
10^{3}	1 000	kilo	k
10^{2}	1 00	hecto	h
10^{1}	10	deca	da
	1 unit		
10^{-1}	0.1	deci	d
10^{-2}	0.01	centi	c
10^{-3}	0.001	milli	m
10^{-6}	0.000 001	micro	m
10^{-9}	0.000 000 001	nano	n
10^{-12}	0.000 000 000 001	pico	p
10^{-15}	0.000 000 000 000 001	femto	f
10^{-18}	0.000 000 000 000 000 001	atto	a
10^{-21}	0.000 000 000 000 000 000 001	zepto	z
10^{-24}	0.000 000 000 000 000 000 000 001	yocto	y

Further reading

Further detail about some of the themes in this book can be found in the following books:

Conventions of technical writing (abbreviations, punctuation, units, etc): physical and life sciences
Scientific Style and Format: the CBE manual for authors, editors and publishers (1994). Sixth edition, Cambridge University Press, Cambridge.

Illustrations
Milne, P H (1992) *Presentation Graphics for Engineering, Science and Business.* Spon, London.

Journal abbreviations: one of the most convenient publications is: *Periodical Title Abbreviations* (1988) Volumes 1-3, edited by L.G. Alkire, and published by Gale Research Company, Detroit, Michigan.

Revising
Venolia, J. (1987) *Rewrite Right! How to Revise Your Way to Better Writing.* Ten Speed Press/Periwinkle Press, Berkeley, California.

Simple statistics and error analysis
Burghardt, M. D. (1992) *Introduction to Engineering*. Chapter 7, 158-181. HarperCollins Publishers Inc., New York.

Writing interpretive essays
Peters, P. *Strategies for Student Writers: A Guide to Writing Essays, Tutorial Papers, Exam Papers and Reports* (1985) John Wiley and Sons, Brisbane.

Index